1007424375

The Role of Science Teachers' Beliefs in International Classrooms

The Role of Science Teachers' Beliefs in International Classrooms

From Teacher Actions to Student Learning

Edited by

Robert Evans
University of Copenhagen, Denmark

Julie Luft
University of Georgia, USA

Charlene Czerniak
University of Toledo, USA

and

Celestine Pea
National Science Foundation, Arlington, USA

SENSE PUBLISHERS
ROTTERDAM/BOSTON/TAIPEI

A C.I.P. record for this book is available from the Library of Congress.

ISBN: 978-94-6209-555-7 (paperback)
ISBN: 978-94-6209-556-4 (hardback)
ISBN: 978-94-6209-557-1 (e-book)

Published by: Sense Publishers,
P.O. Box 21858,
3001 AW Rotterdam,
The Netherlands
https://www.sensepublishers.com/

Printed on acid-free paper

TABLE OF CONTENTS

TABLE OF CONTENTS

SECTION ONE

INTRODUCTION: OVERVIEW OF THE FIELD

RON BLONDER, NAAMA BENNY & M. GAIL JONES

TEACHING SELF-EFFICACY OF SCIENCE TEACHERS

INTRODUCTION

Whether one examines teachers' effectiveness from the perspective of a legislator, parent, principal, or student, the main goal is to prepare teachers who have a strong knowledge base related to science, knowledge of effective teaching strategies, the ability to teach, and a desire to make a difference in the lives of their students. The underlying construct that influences each of these factors is teachers' self-efficacy. We now know that teachers' self-efficacy is embedded in an integrated system that includes prior experiences including previous successes and failures, and feedback from others. Self-efficacy can shape how a teacher will implement a new curriculum, predict the success or failure of a textbook or other curricular material, influence the effectiveness of professional development, or effectively frame a teacher's response to a student's question. Giving tribute to Bandura's work (1986), Gibbs (2002) noted that teachers' self-efficacy beliefs are "suggested as impacting on how teachers think, feel, and teach." (p. 1). In this chapter we define and describe the elusive construct of teachers' self-efficacy and what specific sources influence teachers' self-efficacy. The relationship of self-efficacy within the larger context of attitudes and beliefs is examined.

We will also review the array of assessments of self-efficacy for teachers of mathematics and science and describe specific studies of teachers' self-efficacy. Finally, we outline future studies that are needed to more fully understand the interacting system that influences teachers' beliefs and self-efficacy. Research has shown that teachers' attitudes and beliefs are embedded in an overlapping network of belief systems (Jones & Carter, 2007). Keys and Bryan (2001) argued that every component of teaching is shaped by teachers' attitudes and beliefs. Furthermore, researchers maintain that new knowledge about teaching and learning is constructed relative to these existing networks of beliefs (Putnam & Borko, 1997). We will examine ongoing research on beliefs within the overarching framework of self-efficacy. The goal is to illustrate the power of teachers' self-efficacy in shaping and defining their practice and to point out directions for future research in the field.

SELF-EFFICACY: SOURCES OF INFLUENCE

Major Influence on Self-efficacy

Bandura (1997) defined self-efficacy as, "[P]eople's beliefs about their capabilities to produce designated levels of performance that exercise influence over events that

R. Evans et al., (Eds.), The Role of Science Teachers' Beliefs in International Classrooms, 3–16.
© *2014 Sense Publishers. All rights reserved.*

affect their lives. (p. 71). Although there are many factors that influence human behavior, Bandura identified self-efficacy as a key mechanism that influences both task performance and the cultivation of cognitive skills. A person's cognitive processes are affected by his or her perceived self-efficacy. A high self-efficacy fosters aspirations for challenging intrinsic goals and promotes good analytical thinking. Negative beliefs about one's abilities can lead to erratic analytical thinking and consequently, the quality of performance deteriorates. Bandura posits that there are four main sources that influence efficacy: mastery experience, vicarious experience, verbal persuasion, and emotional arousal (Bandura & Adams, 1977; Bandura, 1982; 1986; 1993; 1997).

Mastery Experience. This can be defined as the interpreted result of one's own previous accomplishments. After completing a task, people interpret and evaluate the results obtained from performing the task. Then, judgment of competence is created or revised according to those interpretations (Usher & Pajares, 2008). Successes build a strong personal efficacy belief, and failures undermine it, especially if failures occur before the self-efficacy belief of a person is firmly established (Bandura, 1994; Usher & Pajares, 2008). If people experience only easy successes they may expect quick results and may be easily discouraged by failure. Developing resilient self-efficacy beliefs requires experience in overcoming obstacles through continuous effort (Bandura, 1994; Usher & Pajares, 2008). However, the amount of effort required to accomplish a task can also indicate the person's ability level. If a person experiences failure after investing a lot of effort, his self-efficacy beliefs may be undermined. Similarly, success that can be achieved only with the help of others provides a weaker sense of the persons' abilities and strengths (Usher & Pajares, 2008). Therefore, when a teacher experiences success helping a difficult student learn, the teacher's self-efficacy beliefs will be influenced. Facing success along with challenging students may have a greater positive influence on a teacher's self-efficacy beliefs than successful experiences with students without difficulties. Some of the most powerful influences on the development of teacher efficacy are mastery experiences during student teaching and during the induction year (Hoy & Woolfolk, 1990).

Vicarious Experiences. Creating and strengthening self-efficacy beliefs by observing others, which one perceives as similar, yields the most comparative information (Usher & Pajares, 2008). Observing people who have undergone similar experiences raises observers' beliefs that may can also possess the capabilities needed to master similar tasks and to succeed (Bandura, 1994; 1997; Usher & Pajares, 2008). The impact of modeling is strongly influenced by observing the experiences of others, especially when the observers are uncertain about their own abilities or have limited previous experience in specific areas. The more that the models are similar to the observers, the better they predict successes, and conversely, failures for the observers. If people perceive the models as being very different from themselves, their perceived self-efficacy belief is not appreciably influenced by the model's behavior and actions. People seek proficient models that hold the competencies to which they aspire. These models can influence their behavior, thinking, the way they transform knowledge and the strategies used for managing environmental demands. Learning from a more capable model such as a master teacher can raise self-efficacy beliefs (Bandura, 1994; Usher & Pajares, 2008). In a study of chemistry teachers, Blonder, Jonatan, Bar-Dov, Benny, Rap, and Sakhnini, (2013) found that vicarious experience was an effective influence on teachers' self-efficacy beliefs regarding the adoption of new instructional technology skills. Teachers noted, that their peers were effective models that taught them how to use new technologies.

Verbal and social persuasion. If people can be convinced verbally by others that they possess the capabilities needed to master a given task, they are likely to invest greater effort and sustain it when problems arise (Bandura, 1994; Usher & Pajares, 2008). The more they are convinced of their inherent ability, the more they will make an effort and try to develop the necessary skills to accomplish a given task. By doing so, they strengthen their self-efficacy beliefs. It is more difficult to construct high self-efficacy beliefs solely through social persuasion than it is to diminish them (Bandura, 1994; 1997; Usher & Pajares, 2008; Zimmerman, 2000). Unrealistic praise or success can lead to disappointing results, and subsequently lead to lowered self-efficacy beliefs. On the other hand, people who are convinced that they lack the necessary capabilities tend to avoid challenging tasks and give up quickly in the face of difficulties (Bandura, 1994). Effective mentors (successful self-efficacy builders) can promote a positive sense of efficacy by structuring situations that bring about success and enabling people to avoid situations in which they are likely to fail. Good mentors measure success in terms of self-improvement rather than by triumphs over others (Bandura, 1994; Usher & Pajares, 2008). The importance of such mentors is particularly important in the first years of teaching where can lead to support increased efficacy beliefs (Woolfolk Hoy, 2000).

Emotional and physiological states. People rely on their physical and emotional states such as anxiety, stress, fatigue and mood in judging their capabilities (Bandura, 1994; Usher & Pajares, 2008). They tend to interpret their physiological arousal as an indicator of personal competence. Stress reactions and tension can be interpreted as signs of vulnerability to poor performance. Mood also affects people's judgments of their self-efficacy. A positive mood enhances a sense of self-efficacy; a pessimistic and low mood diminishes it. It is not the emotional and physical states themselves that are important but rather, how they are perceived and interpreted (Bandura, 1994; 1997; Usher & Pajares, 2008). In general, according to Bandura (1997), increasing physical and emotional well-being and reducing negative emotional states strengthen self-efficacy beliefs.

Many factors influence the way teachers weigh, interpret, and integrate information from these four sources in evaluating their own teaching capabilities. The outcomes teachers expect largely depend on their judgment of how well they will be able to perform in given situations. Their judgment is a context-specific assessment of their competence to do something specific. They visualize outcomes and then deduce their own capabilities to perform specific teaching tasks (Bandura, 1997; Pajares, 1997; Woolfolk Hoy, Hoy, & Davis, 2009). The most effective way of creating a strong sense of efficacy is by undergoing mastery experiences: the interpreted result of one's own previous accomplishments. Mastery experiences consistently emerge in empirical studies as the most powerful source of self-efficacy across domains (Usher & Pajares, 2008; Zimmerman, 2000).

TEACHING SELF-EFFICACY DEFINITIONS AND ITS IMPORTANCE

Tschannen-Moran, Woolfolk Hoy, and Hoy (1998) defined teaching self-efficacy as "teacher's belief in her or his ability to organize and execute the courses of action required to successfully accomplish a specific teaching task in a particular context" (p. 22). According to Gibson and Dembo (1984), "Teachers who believe students' learning can be influenced by effective teaching (outcome expectancy beliefs) and also have confidence in their own teaching abilities (self-efficacy beliefs) should persist longer, provide a greater focus in the classroom, and exhibit different types of feedback than teachers who have lower expectations concerning their ability to influence student learning" (p. 570). This definition includes teachers' judgment of their ability to bring about desired outcomes of students' engagement and learning, even among those students who may be difficult or who are unmotivated. Friedman and Kass (2002) expanded the definition of teaching efficacy. They referred to teacher self-efficacy as a two-factor concept, embracing two interrelated efficacies: classroom efficacy and organizational efficacy. "[T]eacher's perception of his or her ability to (a) perform required professional tasks and to regulate relations involved in the process of teaching and educating students (classroom efficacy), and (b) perform organizational tasks, become part of the organization and its political and social processes (organizational efficacy)" (p. 684).

Teaching is a process that includes several components: The teacher herself, students within her classroom, the content, and the school environment. Table 1 examines teaching efficacy in light of these components.

Tschannen-Moran et al. (1998) presented a model for teacher efficacy. According to this model teachers' efficacy judgments result from analyses of teaching tasks and assessments of their personal teaching competence. This analysis takes place in the context of personal assessment of those factors that make a specific task easy or difficult. These researchers maintain that task judgments are specific to an individual teacher's self assessment of his or her personal teaching capabilities and limitations that are specific to the teaching task.

Although defined differently, most researchers agree that teachers' efficacy is positively related to important and desired teaching consequences. Teaching self-efficacy positively affects students' outcomes such as achievement, motivation, and

Table 1. Components of teaching self-efficacy

Leading Question	Dimension	Definition
Which factors are influenced by teaching self-efficacy? How does teaching self-efficacy influence student learning?	Teachers' self-efficacy beliefs affect not only the teacher but also the level of academic achievements and the motivation to learn of his/her students.	Bandura (1993): "Teachers' beliefs in their personal efficacy to motivate and promote learning affect the types of learning environments they create and the level of academic progress their students achieve" (p. 117) Berman, McLaughlin, Bass, Pauly & Zellman, (1977). (In: Tschannen-Moran et al., 1998): "The extent to which the teacher believes he or she has the capability to affect student performance" (p. 137) Guskey and Passaro (1994): "Teachers' beliefs or conviction . . . can influence how well students learn, even those who may be difficult or unmotivated" (p. 4)
How does teaching self-efficacy influence and shape teachers' professionalization?	Teachers' beliefs have consequences for their behavior and therefore on the amount of effort they exert. Teacher's efficacy beliefs that include both personal competence and the analysis of the task influence their teaching professional behavior.	Bandura (1997): "influence the courses of action people choose to pursue, how much effort they put forth in given endeavors, how long they will persevere in the face of obstacles and failures, their resilience to adversity" (p.3) Gibson and Dembo (1984): "Teachers who believe students' learning can be influenced by effective teaching (outcome expectancy beliefs) and also have confidence in their own teaching abilities (self-efficacy beliefs) should persist longer, provide a greater focus in the classroom, and exhibit different types of feedback than teachers who have lower expectations concerning their ability to influence student learning" (p. 570) Woolfolk Hoy et al. (2009) "Teachers' efficacy judgments are the result Woolfolk Hoy et al. (2009) of an interaction between (a) a personal appraisal of the factors that make accomplishing a specific teaching task easy of difficult (analysis of teaching task in context) and (b) a self-assessment of personal teaching capabilities and limitations specific to the task (analysis of teaching competence)." (p. 628)

(Continued)

Table 1. Components of teaching self-efficacy (Continued)

Leading Question	Dimension	Definition
Is teaching self-efficacy domain-specific or domain-general?	A teacher's efficacy belief can vary by teaching subjects. A teacher can have high sense of efficacy for one subject but at the same time have a moderate or low sense of efficacy for other subject	Woolfolk Hoy et al. (2009): " Woolfolk Hoy et al. (2009): One of the unresolved issues in the measurement of teacher efficacy is determining the optimal level of specificity. For example, is efficacy specific to teaching mathematics, or more specific to teaching algebra, or even more specific to teaching quadratic equations?" (p. 631)

Woolfolk Hoy et al. (2009): "[T]eachers who lack confidence in their knowledge of science content and pedagogy tend to deemphasize or avoid science teaching or teach using transmissive as opposed to inquiry methods" (p. 632).

Tschannen-Moran, Woolfolk Hoy and Hoy (1998): "A teacher may feel very competent in one area of study or when working with one kind of student and feel less able in other subject or with different students" (p. 10) |
| What is the relationship of teaching efficacy and the larger school context? | Teachers' self-efficacy is not only teaching efficacy but also has a dimension of school organization efficacy. | Friedman and Kass (2002): "[T]eacher's perception of his or her ability to (a) perform required professional tasks and to regulate relations involved in the process of teaching and educating students (classroom efficacy), and (b) perform organizational tasks, become part of the organization and its political and social processes (Organizational efficacy)." (p.684) |

the students' own sense of efficacy (Ashton & Webb, 1986; Cakiroglu, Capa-Aydin, & Woolfolk Hoy, 2012; Guskey, 1981; 1988; Lumpe, Czerniak, Haney, Beltyukova, 2011; Pajares, 1993; Tschannen-Moran et al., 1998; Tschannen-Moran & Woolfolk Hoy, 2001). In addition, teachers' efficacy beliefs are related to classroom behavior, and the effort that teachers invest in teaching. Teachers with a strong sense of efficacy tend to be more organized and generally plan better than those without a strong sense of efficacy. They also tend to be more open to new ideas and innovations, more willing to experiment with new teaching methods, are better in meeting the needs of their students, and are more likely to use powerful but potentially difficult-to-manage methods such as inquiry and small-group work (Ashton & Webb, 1986;

8

Guskey & Passaro, 1994; Pajares, 1993; Tschannen-Moran et al., 1998; Tschannen-Moran & Woolfolk Hoy, 2001). Marshall, Horton, Igo, and Switzer (2008) studied over a thousand teachers at the elementary and secondary levels and found that teachers with higher self-efficacy were more likely to have their students engage in inquiry. Other research has shown that greater self-efficacy beliefs empower teachers to be less critical regarding students' mistakes, to work longer with students who are struggling (Ashton & Webb, 1986), and to exhibit greater enthusiasm and commitment to teaching (Tschannen-Moran & Woolfolk Hoy, 2001).

Since the concept of teaching self-efficacy has been found to strongly influence so many desired leaning outcomes, different scales and instruments for teaching self-efficacy have been developed (Henson, Kogan, & Vacha-Haase, 2001). One of the early assessments to measure personal teaching efficacy (PTE) and general teaching efficacy (GTE) was developed by Gibson and Dembo (1984). Riggs and Enochs (1990) developed an elementary science teacher efficacy belief scale that measures personal science teaching efficacy (PSTE) and science teaching outcome expectancy (STOE). Tschannen-Moran and Woolfolk Hoy (2001) developed the Ohio Teacher Sense of Efficacy Scale, according to their model for teaching self-efficacy. This scale measures student engagement, instructional practices, and classroom management, and was validated in different teaching settings Woolfolk Hoy et al., 2009 and in different countries Klassen et al., 2009. Different measurement scales for teaching self-efficacy are discussed in section two of this book (see Chapters 3 and 4).

SCIENCE TEACHERS' SELF-EFFICACY: GENERAL RESEARCH TRENDS

There are limited numbers of studies that have explored the area of teaching self-efficacy with science teachers. Of the studies that exist, there are two primary lines of research. The first group of research studies was conducted during the nineties (e.g., Cannon, Scharmann, 1996; Ramey-Gassert, Shroyer, & Staver, 1996), and renewed interest in the field has emerged in recent years (Gunning, Mensah, 2011; Hechter, 2011; Boone, Townsend, Staver, 2011; Posnanski, 2002; Bleicher, Lindgren, 2005; Azar, 2010). Most of the research on science teaching self-efficacy has focused on elementary science teachers, which was measured with quantitative research methods, with only a few exceptions. Azar (2010) explored high-school science-teaching self-efficacy and Ramey-Gassert et al. (1996) applied qualitative research methodology to identify the factors that influence science teaching self-efficacy of elementary school teachers.

Wheatly (2005) criticized the quantitative research of teaching efficacy and raised questions regarding the practical use of this research in teacher education "[W]hy isn't it clearer how to use teacher efficacy research in teacher education? Teacher efficacy researchers have conducted many well-crafted studies and are interested in practical applications of their work. Why aren't the practical fruits of their research more obvious?" (p. 748). These questions are valid in the field of science teaching as well. In fact, there have been some attempts to apply teaching efficacy to pre-service science teachers' programs and to in-service professional development programs of science teachers.

Science teachers, like all teachers, possess beliefs about teaching and learning that influence their behavior and practice. Understanding the nature of science and how students learn science aid in forming a set of beliefs that guide practice and behavior within the classroom (Bryan, 2012; Riggs & Enochs, 1990). Teachers' beliefs about themselves and about their students as science knowers have also influenced behavior and practice (Laplante, 1997). When teachers view themselves as consumers of science knowledge, and view science as a body of knowledge, their teaching reflects these beliefs – they use more teacher-centered strategies in which knowledge is controlled and transmitted by the teacher. When the teachers hold a social constructivist view about science, they are more willing to use open-ended science inquiry projects (Bryan, 2012; Woolfolk Hoy et al., 2009). Evans (2011) evaluated a professional development program for science teachers aimed at developing inquiry-based teaching. In this workshop, active strategies to enhance teaching efficacy were used. These strategies resulted in a long-term increase in teachers' efficacy. Teaching efficacy research has been used to evaluate other in-service science teachers' professional development programs. Lakshmanan, Heath, Perlmutter, and Elder (2011) examined the impact of standards-based professional development on teacher efficacy and instructional practice of elementary and middle school science teachers and found significant growth in teacher self-efficacy but not in outcome expectancy. There was also significant growth in the extent to which teachers implemented inquiry-based instruction in the classroom and a positive correlation was observed between changes in self-efficacy and changes in the use of inquiry-based instructional practice. Khourey-Bowers and Simonis (2004) analyzed the influence of program design on achieving gains in personal science teaching self-efficacy, outcome expectancy, chemistry content, and pedagogical content knowledge in a three-year professional development program for chemistry middle-school teachers. They concluded that "Significant changes in *both* subscales of the Science Teaching Efficacy Belief Instrument (STEBI), though not usually realized, are necessary professional development outcomes if national science education goals are to be achieved. Teachers' attitudes toward teaching science affect choices they make in classroom content and strategy." (p. 193). In addition, the role of science teachers' beliefs is significantly related to how they implement the curriculum (Bencze, Bowen, & Alsop, 2006; Laplante, 1997).

Science teachers' content knowledge plays an important role in their science teaching efficacy beliefs (Palmer, 2006; Posnanski, 2002). Palmer (2006) suggested that science content knowledge (CK) is a type of mastery experience. This type of mastery experience involves success in understanding something rather than success in doing something. It could therefore be referred to as "cognitive content mastery" (Palmer, 2006, p. 339). Cognitive content mastery was a source of self-efficacy for 9%-19% of students in an academic teachers' education course (Palmer, 2006). Strong science content knowledge, in association with teaching methods, constitutes the foundation of efficacious science teaching. Effective science knowledge also helps establish science teachers' self-efficacy by reducing anxiety about science and science teaching (Bryan, 2012; Palmer, 2006; Posnanski, 2002).

Teachers' educational experiences may have powerful effects on their self-efficacy beliefs (Smolleck & Mongan, 2011). Ramey-Gassert et al. (1996) and Coble and Koballa (1996) indicated that teachers' attitudes toward teaching science, and as a result, their self-efficacy beliefs, can be altered through training sessions that focus on both learning and teaching science context. Professional development programs typically promote the development of knowledge about science content, student learning, and teaching methods. Researchers maintain that to change teachers' beliefs and behaviors professional development programs need to be of sufficient duration (Loucks-Horsley, Hewson, Love, & Stiles, 1998; Posnanski, 2002; Reys, Reys, Barnes, Beem, & Papik, 1997). Loucks-Horsley et al. (1998) identified the critical need for content information to be embedded within professional development experiences for science teachers. Posnanski's study (2002) indicated that a professional development program built on a research-based model could contribute to positive changes in science teachers' self-efficacy beliefs and potentially change their teaching behavior.

CONCLUSION AND FUTURE RESEARCH

In this chapter we defined and described the construct of teachers' self-efficacy and what specific sources influence teachers' self-efficacy. We followed the development of the concept of teaching self-efficacy and described relevant studies that focus on teachers' self-efficacy. Finally, we have suggested future studies that might provide a deeper understanding of the interacting systems that influence teachers' beliefs and self-efficacy.

Based on a review of the existing literature, there are a number of areas where further research is needed. There is a need to move beyond the quantitative surveys of teaching efficacy to develop a better understanding of the contextualized pathways and experiences that promote teachers' development of teaching efficacy. Wheatley (2005) raised serious doubts about the existing teaching efficacy scales: "[E]xisting scales, yielding global and numerical levels of "teacher efficacy", do not reveal what teachers' responses mean, or where they need support from teacher educators." (p. 751). In order to address those issues, we need focused studies that can unpack these factors. For example, the Blonder, et al, (2013) study of chemistry teachers' self-efficacy beliefs showed that teacher professional development can build on teachers' self efficacy and promote growth. But simple quantitative measures of self-efficacy may not be sufficient and more rich detailed measures of self-efficacy are needed.

Additional research is needed to further refine and define the construct of teaching self efficacy and the factors that influence this construct. Are there components of teaching efficacy such as organizational efficacy as suggested by Friedman and Kass (2002)? To what degree does organizational efficacy contribute to overall teaching self-efficacy? We need to know more about how teaching efficacy may differ across contexts and student characteristics. A teacher with high teaching efficacy beliefs has the ability to teach even the most difficult students in the class (Gibson & Dembo, 1984; Tschannen-Moran and Woolfolk Hoy, 2001). However, little is known about teachers' efficacies in teaching students with diverse needs such as those with learning disabilities or those identified as gifted and talented. In one of the few studies that exist on this topic, Starko and Schack (1989) investigated teacher efficacy and its effect on how different teaching strategies used for teaching gifted students. It was found that teachers respond differently to various students' characteristics (e.g., Touranki, 2003; Tournaki & Podell, 2005), and that their sense of efficacy influences their academic predictions, in relation to different student characteristics. More research in the field of teaching efficacy in teaching students with special learning characteristics is needed, especially for science teachers (Taber, 2007; Van Tassel-Baska, Bass, Reis, Poland, & Avery, 1998). What is the relationship between teachers' teaching efficacy and students' with special needs self -efficacy in learning science?

In this chapter we reviewed both historic and the latest contributions in the field of science teaching efficacy. We described studies about elementary and middle school science teaching efficacy. We wish to stress the importance of conducting research beyond the elementary school level, where more complex science subject matter is taught. We know that secondary school students' interactions with their teachers can affect their future choices (Sjaastad, 2011). How do high efficacy science teachers influence their students' future career? And how do interactions between students and their science teachers affect the teachers' teaching efficacy?

For teacher educators there are a number of unanswered questions related to teaching self-efficacy that should be explored to maximize the impact of the teacher education program. We need to know more about the developmental trajectory that takes place as teachers develop self-efficacy. What types of experiences are most effective in promoting the development of teaching self-efficacy in beginning teachers? How can mentoring promote positive self-efficacy that allows teachers to overcome obstacles? Can virtual reality situations (virtual vicarious experiences) simulate classroom experiences and influence the development of self-efficacy? If emotional stress reduces self-efficacy beliefs as suggested by research (Bandura, 1997) is it possible to provide effective mentoring experiences to ameliorate the negative influence of stress? Are there ways that teacher education programs can promote the development of teaching self-efficacy as a result of group experiences? How is teaching self-efficacy context dependent? Which components of teaching self-efficacy are carried across contexts? If teacher educators can promote teaching self-efficacy can we promote the likelihood that a teacher will use inquiry-based methods in science instruction? How does science content knowledge contribute to teaching self-efficacy related to teaching science process skills or the nature of science?

These questions and others regarding science teaching efficacy should be addressed in future research. Unraveling the factors, contexts, and influences to promote positive teaching self-efficacy has the potential to payoff in terms of more effective instruction that promotes deep understandings of science content and processes.

REFERENCES

Ashton, P. T., & Webb, R. B. (1986). *Making a difference: Teachers' sense of efficacy and student achievement.* New York, NY: Longman.

Azar, A. (2010). In-service and pre-service secondary science teachers' self-efficacy beliefs about science teaching. *Educational Research and Reviews, 5,* 175–188.

Bandura, A. (1982). Self–efficacy mechanism in human agency. *American Psychology, 37,* 122–147.

Bandura, A. (1986). The explanatory and predictive scope of self-efficacy theory. *Journal of Social and Clinical Psychology, 4,* 359–373.

Bandura, A. (1993). Perceived self-efficacy in cognitive development and functioning. *Educational Psychologist, 28,* 117–148.

Bandura, A. (1994). Self-efficacy. In V. S. Ramachaudran (Ed.), *Encyclopedia of human behavior* (Vol. 4, pp. 71–81). New York, NY: Academic Press.

Bandura, A. (1997). *Self-efficacy: The exercise of control.* New York, NY: Freeman.

Bandura, A., & Adams, N. E. (1977). Analysis of self-efficacy theory of behavioral change. *Cognitive Therapy and Research, 1,* 287–310.

Bencze, J. L., Bowen, G. M., & Alsop, S. (2006). Teachers' tendencies to promote student-led science projects: Associations with their views about science. *Science Education, 90,* 400–419. doi: 10.1002/sce.20124

Bleicher, R., & Lindgren, J. (2005). Success in science learning and preservice science teaching self-efficacy. *Journal of Science Teacher Education, 16,* 205–225. Doi:10.1007/s10972-005-4861-1

Blonder, R., Jonatan, M., Bar-Dov, Z., Benny, N., Rap, S., & Sakhnini, S. (2013). Can You Tube it? Providing chemistry teachers with technological tools and enhancing their efficacy beliefs. *Chemistry Education Research and Practice, 14,* 269–285. Doi:10.1039/c3rp00001j

Boone, W. J., Townsend, J. S., & Staver, J. (2011). Using Rasch theory to guide the practice of survey development and survey data analysis in science education and to inform science reform efforts: An exemplar utilizing STEBI self-efficacy data. *Science Education, 95,* 258–280. Doi:10.1002/sce.20413

Bryan, L. A. (2012). Research on science teacher beliefs. In B. J. T. Fraser, K. McRobbie, J. Campbell, & J. Barry (Eds.), *Second international handbook of science education* (Vol. 24, pp. 477–495). New York, NY: Springer.

Cakiroglu, J., Capa-Aydin, Y., & Woolfolk Hoy, A. (2012). Science teaching efficacy beliefs. In B. J. Fraser, K. Tobbin, & C. J. McRobbie (Eds.), *Second international handbook of science education* (Vol. 24, pp. 477–495). New York, NY: Springer.

Cannon, J. R., & Scharmann, L. C. (1996). Influence of a cooperative early field experience on preservice elementary teachers' science self-efficacy. *Science Education, 80,* 419–436. doi:10.1002/(sici)1098-237x(199607)80:4<419::aid-sce3>3.0.co;2-g

Evans, R. H. (2011, September). *Active strategies during inquiry-based science teacher education to improve long-term teacher self-efficacy.* Paper presented at the European Science Education Research Assocation, Lyon, France.

Friedman, I. A., & Kass, E. (2002). Teacher self-efficacy: A classroom-organization conceptualization. *Teaching and Teacher Education, 18,* 675–686. Doi:10.1016/s0742-051x(02)00027-6

Gibbs, C. (2002, September). *Effective teaching: Exercising self-efficacy and thought control of action.* Paper presented at the Conference of the British Educational Research Association, Exeter, United Kingdom. Retrieved from http://www.leeds.ac.uk/educol/documents/00002390.htm

Gibson, S., & Dembo, M. H. (1984). Teacher efficacy: A construct validation. *Journal of Educational Psychology, 76,* 569–582.

Gunning, A., & Mensah, F. (2011). Preservice elementary teachers' development of self-efficacy and confidence to teach science: A case study. *Journal of Science Teacher Education, 22,* 171–185. doi:10.1007/s10972-010-9198-8

13

Guskey, T. R. (1981). Measurement of the responsibility teachers assume for academic successes and failures in the classroom. *Journal of Teacher Education, 32*, 44–51. doi:10.1177/002248718103200310

Guskey, T. R. (1988). Teacher efficacy, self-concept, and attitudes toward the implementation of instructional innovation. *Teaching and Teacher Education, 4*, 63–69. doi:10.1016/0742-051x(88)90025-x

Guskey, T. R., & Passaro, P. D. (1994). Teacher efficacy: A study of construct dimensions. *American Educational Research Journal, 31*, 627–643. doi:10.3102/00028312031003627

Hechter, R. (2011). Changes in preservice elementary teachers' personal science teaching efficacy and science teaching outcome expectancies: The influence of context. *Journal of Science Teacher Education, 22*, 187–202. doi: 10.1007/s10972-010-9199-7

Henson, R. K., Kogan, L. R., & Vacha-Haase, T. (2001). A reliability generalization study of the teacher efficacy scale and related instruments. *Educational and Psychological Measurement, 61*, 404–420.

Hoy, A.W., Hoy, W. K., & Davis, H. (2009). Teachers' self-efficacy beliefs. In K. Wentzel & A. Wigfield (Eds.), Handbook of motivation in school (pp. 627–653). Mahwah, NJ: Erlbaum.

Jones, M. G., & Carter, G. (2007). Science teacher attitudes and beliefs. In S. K. Abell & N. G. Lederman (Eds.), *Handbook of research on science education* (pp. 1067–1104). Mahwah, NJ: Lawrence Erlbaum Associates.

Khourey-Bowers, C., & Simonis, D. (2004). Longitudinal study of middle grades chemistry professional development: Enhancement of personal science teaching self-efficacy and outcome expectancy. *Journal of Science Teacher Education, 15*, 175–195.

Klassen, R. M., Bong, M., Usher, E. L., Chong, W. H., Huan, V. S., Wong, I. Y. F., & Georgiou, T. (2009). Exploring the validity of a teachers' self-efficacy scale in five countries. *Contemporary Educational Psychology, 34*, 67-76. doi: http://dx.doi.org/10.1016/j.cedpsych.2008.08.001

Keys, C. W., & Bryan, L. A. (2001). Co-constructing inquiry-based science with teachers: Essential research for lasting reform. *Journal of Research in Science Teaching, 38*, 631–645.

Lakshmanan, A., Heath, B. P., Perlmutter, A., & Elder, M. l. (2011). The impact of science content and professional learning communities on science teaching efficacy and standards-based instruction. *Journal of Research in Science Teaching, 48*(5), 534–551. doi:10.1002/tea.20404

Laplante, B. (1997). Teachers' beliefs and instructional strategies in science: Pushing analysis further. *Science Education, 81*, 277–294. Doi:10.1002/(sici)1098-37x(199706)81:3<277::aid-ce2>3.0.co;2-a

Loucks-Horsley, S., Hewson, P. W., Love, N., & Stiles, K. E. (1998). *Designing professional development for teachers of science and mathematics.* Thousand Oaks, CA: Corwin Press.

Marshall, J., Horton, J., Igo, B., & Switzer, D. (2009). K-12 science and mathematics teachers beliefs about and use of inquiry in the classroom. *International Journal of Science and Mathematics Education, 7*, 575–596.

Pajares, F. (1993). Preservice teachers' beliefs: A focus for teacher education. *Action in Teacher Education, 15*, 45–54.

Pajares, F. (1997). Current directions in self-efficacy rResearch. In M. L. Maehr & P. R. Pintrich (Eds.), *Advances in motivation and achievement* (Vol. 10, pp. 1–49). Greenwich, CT: JAI Press.

Palmer, D. (2006). Sources of self-efficacy in a science methods course for primary teacher education students. *Research in Science Education, 36*, 337–353. Doi:10.1007/s11165-005-9007-0

Posnanski, T. J. (2002). Professional development programs for elementary science teachers: An analysis of teacher self-efficacy beliefs and a professional development model. *Journal of Science Teacher Education, 13*, 189–220. Doi:10.1023/a:1016517100186

Putnam, R., & Borko, H. (1997). Teacher learning: Implications of new views of cognition. In B. J. Biddle, T. L. Good, & I. F. Goodson (Eds.), *International handbook of teachers and teaching* (pp. 1223–1296). Amsterdam, The Netherlands: Kluwer.

Ramey-Gassert, L., Shroyer, M. G., & Staver, J. R. (1996). A qualitative study of factors influencing science teaching self-efficacy of elementary level teachers. *Science Education, 80*, 283–315. Doi:10.1002/(sici)1098-237x(199606)80:3<283::aid-sce2>3.0.co;2-a

Reys, B. J., Reys, R. E., Barnes, D., Beem, J., & Papik, I. (1997). Collaborative curriculum investigations as a vehicle for teacher enhancement and mathematicscurriculum reform. *School Science and Mathematics, 97*, 422–426.

Riggs, I., & Enochs, L. (1990). Toward the development of an elementary teacher's science teaching efficacy belief instrument. *Science Education, 74*, 625–638.

Sjaastad, J. (2011). Sources of inspiration: The role of significant persons in young people's choice of science in higher education. *International Journal of Science Education, 34*(10), 1–22. Doi:10.1080/09500693.2011.590543

Smolleck, L. A., & Mongan, A. M. (2011). Changes in preservice teachers' self-efficacy: From science methods to student teaching. *Journal of Educational and Developmental Psychology, 1*, 133–145. Doi:10.5539/jedp.v1n1p133

Starko, A. J., & Schack, G. D. (1989). Perceived need, teacher efficacy, and teaching strategies for the gifted and talented. *Gifted Child Quarterly, 33*, 118–122.

Taber, K. S. (Ed.). (2007). *Science education for gifted learners.* London, UK: Routledge.

Tschannen-Moran, M., Woolfolk Hoy, A., & Hoy, W. K. (1998). Teacher efficacy: Its meaning and measure. *Review of Educational Research, 68*, 202–248.

Tschannen-Moran, M., & Woolfolk Hoy, A. (2001). *Teacher efficacy: capturing an elusive construct. Teaching and Teacher Education, 17*, 783-805. doi: http://dx.doi.org/10.1016/S0742-051X(01)00036-1

Tournaki, N. (2003). The differential effects of teaching addition through strategy instruction versus drill and practice to students with and without learning disabilities. *Journal of Learning Disabilities, 36*, 449–558.

Tournaki, N., & Podell, D. M. (2005). The impact of student characteristics and teacher efficacy on teachers' predictions of students› success. *Teaching and Teacher Education, 21*, 299–314.

Usher, E. L., & Pajares, F. (2008). Sources of self-efficacy in school: Critical review of the literature and future directions. *Review of Educational Research, 78*, 751–796. Doi:10.3102/0034654308321456

VanTassel-Baska, J., Bass, G., Reis, R., Poland, D., & Avery, L. D. (1998). A national study of science curriculum effectiveness with high ability students. *Gifted Child Quarterly, 42*(4), 200–211.

Wheatley, K. F. (2005). The case for reconceptualizing teacher efficacy research. *Teaching and Teacher Education, 21*, 747–766.

Woolfolk Hoy, A. (2000, April). Changes in teacher efficacy during the early years of teaching. Paper presented at the annual meeting of the American Educational Research Association, New Orleans, LA. Retrieved from http://citeseerx.ist.psu.edu/viewdoc/download?doi=10.1.1.183.4309&rep=rep1&type=pdf

Woolfolk Hoy, A., Hoy, W., & Davis, H. (2009). Teachers' self-efficacy beliefs. In K. R. W. A. Wigfield (Ed.), Handbook of motivation at school (pp. 627–654). New York, NY: Routledge.

Zimmerman, J. B. (2000). Self-Efficacy: An essential motive to learn. *Educational Psychology, 25*, 82–91. Doi:10.1006/ceps.1999.1016

AFFILIATION

Ron Blonder
Department of Science Teaching
Weizmann Institute of Science, Israel

Naama Benny
Department of Science Teaching
Weizmann Institute of Science, Israel

M. Gail Jones
Department of Mathematics, Science, & Technology Education
North Carolina State, University, USA

CAROLYN S. WALLACE

OVERVIEW OF THE ROLE OF TEACHER BELIEFS IN SCIENCE EDUCATION

INTRODUCTION TO TEACHER BELIEFS IN SCIENCE EDUCATION

The monumental shift in cognitive science from a behaviorist theory of learning to a constructivist theory of learning, taking place from the late 1960s through the late 1980s, had a profound impact on scholars' understanding of teaching and teacher thinking. The literature on teaching throughout the 1980s increasingly referred to teachers as creative and intelligent professionals who make decisions based on their own knowledge, beliefs and experiences (Richardson, 1996). Scholarly literature from the 1980s onward established the construct of teacher beliefs as a form of cognition that greatly influences what happens in classrooms. Beliefs are most often thought of as views, opinions and principles "not immediately susceptible to rigorous proof" (Dictionary.com, 2012). However, the beliefs construct as applied to science teaching has multiple meanings and is subject to interpretation. One often quoted statement about the significance of teacher beliefs to educational research was made by Pajares (1992), "[many researchers agree that] beliefs are the best indicators of the decisions individuals make throughout their lives."

The definition or specification of teacher beliefs has been a point of controversy across the years in the literature. One of the central points of the debate has been whether and how beliefs differ from knowledge (Pajares, 1992). Some scholars assert that beliefs include an affective or evaluative component not encompassed by the knowledge construct, while others have defined knowledge based on experience more broadly. Interested readers may want to read the following articles that depict the evolution of the beliefs construct over time: Bryan (2012), Green (1971), Nespor (1987), Pajares (1992), and van Driel, Beijard & Verloop (2001). Despite these differences in definitions, it is well accepted by researchers that teacher beliefs have a powerful impact on science teaching and learning. Research over the past three decades has resulted in a set of assumptions about the nature of teacher beliefs that are widely accepted (Bryan, 2012). These include:

1. Beliefs are far more influential than academic knowledge in framing, analyzing and solving problems and making teaching decisions.
2. Some beliefs are more strongly held than others, resulting in "core" and "peripheral" beliefs. An individual's core beliefs may be more resistant to change.

R. Evans et al., (Eds.), The Role of Science Teachers' Beliefs in International Classrooms, 17–31.

3. Beliefs do not exist independently of one another, but are arranged in an ecology, or an "internal architecture" of systems that have psychological importance to the individual.
4. Individuals may have competing belief sets about the same topic.
5. When one belief is changed, it is likely to affect other beliefs throughout the system.
6. Some scholars posit that belief systems occur in "nests" (Bryan, 2003) or sets of beliefs, including core and peripheral beliefs about various principles that are linked or grouped together.

In science education, a landmark study by Munby (1984) solidified the importance of teacher beliefs to practices. Munby recognized that teachers are not likely to be convinced to adopt innovative teaching strategies based solely on scientific evidence from research studies. Rather, teachers will take on the important role of interpreting the innovation and evaluating its efficacy for their particular students. Munby asserted, "Importantly, part of a teacher's context which is evidently significant to adopting research findings or implementing curricula is what a teacher believes . . . (1984, p. 28)." Using repertory grid analysis (see Kelly, 1955) and a series of iterative interviews and observations, Munby concluded that the participant teacher in his study, Ellen, had deep seated beliefs that guided her practice. These included: (a) helping students cope with new information and learn independently; (b) increasing student confidence; and (c) helping students learn concepts in the earth science curriculum which she thought were valuable for their everyday lives. Ellen's orientation to teaching was pragmatic rather than theoretical. Munby concluded that Ellen would review and filter new curriculum innovations for those that were resonate with her core beliefs.

The ideas of Nespor (1987), while not specific to science education, have often been adopted by science educators researching teacher beliefs. Nepor's early work helped establish beliefs as a theoretical construct and asserted that teachers rely on their core belief systems rather than academic knowledge when determining classroom actions. Nespor noted that the rapid pace and ill-structured nature of educational environments promotes decision making based on core affective elements and evaluations rather than step-by-step problem solving. He posited that beliefs are made up of: (a) episodic knowledge, characterized by remembered stories and events; (b) affective elements, such as feelings about students, and (c) "existential presumptions," or beliefs about the existence or nonexistence of categorical entities, such as "brightness," "immaturity," "ability" and "laziness." Nespor views teacher beliefs as an integration of knowledge and feelings built up largely through teaching experience.

Van Driel and colleagues (van Driel, Beijard & Verloop, 2001) published a cognitive framework for science teaching in which they depicted beliefs as a subset of teachers' practical knowledge. Along with beliefs as influential determiners of classroom practice, they characterized teachers' practical knowledge as being action-oriented, personal and context-bound, tacit and integrated. Van Driel and colleagues asserted that beliefs act as a "filter" through which newly acquired information is passed before it is integrated into the knowledge base. The idea of beliefs as a filter for knowledge is similar to Munby's (1984) original assertion that teachers will search for aspects of reform-based practice that are compatible with their core beliefs.

A fourth foundational study on the relationship between teacher beliefs and intentions to implement reform-based teaching (Haney, Czerniak & Lumpe, 1996) employed theory and mathematical modeling from educational psychology. Haney and colleagues posited that the intention to implement reform would be a direct result of teachers' attitudes towards the reform behavior, perceived social norms in their school context, and perceived behavioral control, or an assessment of the obstacles or resources available to carry out the behavior. According to the researchers, an individual's salient beliefs lie behind her attitudes, perceptions of social norms and perceptions of behavioral control and are thus indirectly at the root of the intention to carry out reform-based teaching behavior. Survey results indicated that indeed, "teacher beliefs are significant contributors of behavioral intention" (Haney et al., 1996, p. 985). Teachers' attitudes towards reform were the greatest contributor to the model of planned intentions, while perceived behavioral control contributed moderately and perceived norms contributed very little. Since attitudes towards reform were so important, the authors asserted that developing positive attitudes could be an anchor for achieving reform. They further suggested that feelings of self-efficacy or success with reform-based teaching experiences might foster positive attitudes about reform.

EXPERIENCED SCIENCE TEACHER BELIEFS AND THEIR INFLUENCE ON TEACHING

Experienced science teachers have had years to build their belief systems which tend to be complex, integrated and quite stable (Bryan, 2012; Wallace & Kang, 2004; van Driel, Bulte & Verloop, 2005; Wallace & Priestley, 2011). One area of particular interest to researchers has been whether teachers have beliefs that support or impede the implementation of reform-based instruction. As a representative example, van Driel and colleagues (2005), researched teachers' beliefs about and intentions to implement a reform-based chemistry curriculum in the Netherlands. They found that although many teachers had mixed beliefs, some teachers fell squarely into either traditional or reform-based factions. Across the entire sample, there was roughly equal support for the traditional and reform-based curricula. The authors implied that curriculum structures should be flexible enough to allow for teacher choice in implementing curriculum in accordance with their beliefs, since these are strongly held and likely to remain stable.

Core Teacher Beliefs Guide Practice

Many studies have indicated congruence or close correspondence between teachers' espoused beliefs about reform, whether positive or negative, and their classroom practices (see Bryan, 2012). Historically, when reform-based interventions are at odds with teacher beliefs, teachers either refuse to implement these reforms or do so superficially (Cotton, 2006; Cronin-Jones, 1991; Olson, 1981; Yerrick, Parke & Nugent, 1997). For example, Cronin-Jones (1991) conducted two case studies of middle grades teachers implementing an innovative constructivist-based science curriculum. She found that both teachers held strong beliefs that students of this age group need explicit direction, learn best through repeated drill and practice, and that factual content acquisition is the most important goal of science education. Both teachers converted the curriculum guidelines into more superficial instructional activities that matched their beliefs about students and science education. Cotton (2006) found that Canadian secondary science teachers rejected the implementation of a value-based environmental science curriculum because of their strong beliefs that education should be value-neutral, allowing children to form their own opinions. Yerrick et al. (1997) documented that teachers constructed rational arguments to describe their implementation of a reform-based curriculum when talking to the project researchers, when in fact they continued to teach in traditional ways.

In a few studies, when teachers' beliefs coincided with the philosophy of the reforms, they worked enthusiastically to promote the reforms (Levitt, 2001; Wallace & Priestley, 2011). Levitt studied the practices of 16 elementary teachers and found that overall they supported reform-based science instruction because it resonated with their ideas about the importance of student-centered curricula. A study of experienced secondary teachers in Scotland indicated that when teachers held positive beliefs about the general principles behind a government-led formative assessment initiative, they not only implemented the reform strategies, but invented ones of their own (Wallace & Priestley, 2011).

Experienced teachers hold a variety of views about reform (van Driel et al., 2005) and may assert their own beliefs when these are at odds with school policy or social culture. A study of Scottish college biology teachers (equivalent to the community college level in the U. S., Priestley, Edwards, Priestley & Miller, 2012) showed that teachers' positive beliefs about reform were indeed associated with the commitment to assert personal teacher agency, rather than follow the school policy of teaching for test performance. One participant in their study was content to have his students achieve high test scores and did not particularly value the types of outcomes, such as collaboration and connectedness, associated with constructivist-based teaching. This participant had no desire to implement reform-based strategies and therefore maintained the status quo of traditional instruction supported by the school administration. A second participant, who did value the outcomes associated with constructivist-based learning, implemented more student-centered strategies and thus took a risk in asserting her beliefs in opposition to what was valued by school management.

Some studies, however, have indicated that there are often more complex relationships between beliefs and practices, including mismatches between espoused beliefs and observed instruction. Although this phenomenon is found more commonly with preservice teachers, experienced teachers have also shown incongruence between their stated beliefs about the nature of science or learning epistemologies and their own classroom practices. This may be related to the phenomenon that teachers learn new ideas through professional development that appeal to them, but that these ideas are held more peripherally than their core teaching beliefs. For example, Trumbell and colleagues (Trumbell, Scarano & Bonney, 2006) found that two participant middle school teachers both espoused reform-based tenets about the nature of science, which they learned superficially in their graduate coursework. However, neither teacher enacted these inquiry-based aspects of science in their classrooms early in their study. Similarly, three participant primary teachers in Waters-Adams (2006) study held moderately traditional hypo-deductive views of the nature of science, but taught science largely through stating facts. Interestingly, in both of these qualitative studies, one or more of the teachers began to change their beliefs during the study and they will be referred to again in the section on belief change below.

Tension between Teachers' Reform-based Beliefs and School Policy

The theme that some teachers hold very positive beliefs about reform-based teaching, but feel thwarted from implementing these in school culture is emerging as an area of policy concern in many school subjects, of which science is one example. Top-down educational polices and their accompanying discourses have been documented to interrupt "productive pedagogies" that might have focused on more real-world connections, questioning and investigation (Lingard, 2005). This phenomenon has been explored most thoroughly in the United Kingdom and the United States (Au, 2006; Ball, 2003; Hursh, 2007; Lingard, 2005; Ranson, 2003). Powerful political structures including federal and state laws (for example in the U. S.), district level administrations, building administrations and master teacher managers enforce adherence to teaching the mandated curriculum in preparation for standardized tests. Within this climate, teachers are expected to produce high test scores at the expense of other educational values, such as critical thinking or the deep exploration of concepts.

Some studies have indicated that teachers may hold sophisticated views of the nature of science or learning epistemologies, but do not use them extensively when planning and teaching their students. Wallace and Kang (2004) found that a group of high school science teachers held competing beliefs about inquiry-based science. The teachers held private belief sets that included enthusiastic attitudes towards inquiry-based teaching. The teachers felt that inquiry engaged students, developed problem solving skills and promoted autonomous thinking. The teachers confided to the researchers that they would use inquiry much more if it did not conflict with the mandated curriculum. In more public settings, the teachers espoused their public belief sets, in which they supported other methods of teaching concepts and reinforcing these through verification labs.

Another example of the tension between positive beliefs about reform-based teaching and accountability pressures was documented by Wallace (2013), a veteran science educator who returned to the classroom in 2005-2006 to experience teaching high school biology in the contemporary educational context of mandated testing. She found that teaching science through inquiry-based methods promoted student questioning, divergent thinking and often open-ended learning outcomes. Divergent thinking as an educational goal stood in direct contrast to the overall cultural goal of the school for producing correct and convergent answers on standardized tests. These dual and opposing purposes of science teaching created a difficult context within which Wallace could enact her constructivist-based teaching beliefs.

Rop (2002) also reported on this tension when he studied the discourse in the classroom of a veteran chemistry teacher. While the teacher valued students' inquiry questions, he did not devote classroom time to investigating these questions. Rop (2002) noted that powerful and conflicting pressures come into play in the everyday patterns of classroom discourse. The teacher felt that the students' questions interrupted the flow of discourse necessary to teach the mandated curriculum. Rop asserted that science educators need to take seriously the juxtaposition between management expectations for content coverage and teachers' desire to honor student-centered inquiry (Rop, 2002). Teachers feel real pressure from structural and cultural influences in the ecology of school. Studies of the ways that policies thwart teacher agency, therefore, suggest that having positive beliefs about reform might be necessary, but not sufficient to affect reform-based strategies in the classroom.

In contrast, some earlier studies indicated that experienced science teachers generally held positive beliefs about their self-efficacy and intentions to enact reform-based teaching. These studies indicated that teachers in general believed in their own ability to influence learning outcomes, their personal agency in being able to carry out effective learning and to control many aspects of their teaching context (Lumpe, Haney & Czerniak, 2000). Some reasons for this discrepancy may be that: (a) these earlier studies pre-date strict enforcement by schools of accountability policies, such as the "No Child Left Behind" legislation in the United States; and (b) the idea that self-efficacy beliefs research uses different theoretical paradigms including psychology of the individual, rather than socio-cultural studies which emphasize a teacher's interaction with others. It would be interesting for researchers to repeat some of the studies of teachers' self-efficacy beliefs (Haney, Czerniak & Lumpe, 1996; Lumpey, Haney & Czerniak, 2000; Enochs, Scharmann & Riggs, 1995) in the current political climate of accountability and standardized testing. This line of research might further elucidate the significance of the role of school policy in regards to how teachers may choose to enact or not enact practice consistent with their beliefs.

In summary, experienced science teachers have belief sets that are stable, closely held and resistant to change. Experienced teachers have a wide range of opinions about the value of reform-based instructional strategies. Teachers most often enact science instruction that is aligned with their core beliefs, whether these represent positive or negative attitudes towards reform. Beliefs often act as a cognitive filter for teachers as they select particular aspects of reform-based instruction to implement. In some cases, teachers adopt new ideas about topics such as the nature of science or learning epistemologies, but these are not easily integrated into their core belief sets or teaching practices. The phenomenon of science teachers having positive beliefs about the value of reform-based teaching, but finding it difficult to enact these reforms in school cultures of accountability can be a significant barrier to overcoming traditional practice.

The Possibility for Change in Experienced Teachers' Beliefs

Because teacher beliefs are so crucial to practice and intentions to enact reform-based teaching, some researchers have examined whether professional development opportunities can change teacher beliefs. Researchers in this line of investigation have examined whether professional development classes or on-site activities result in modifications to science teaching beliefs and/or practice. These scholars have tried to explain the complex relationships among learning, teaching and belief change.

In one study (Lavonen, Jauhiainen, Koponen & Kurki-Suonio, 2004), experienced physics teachers in Finland participated in a one and one-half year professional development program that emphasized the use of lab work in physics. The designers of the program sought to change teachers beliefs about the ways labs might be used to foster conceptual development. They sought to develop teachers' ideas that lab work could be used to build students' epistemic knowledge of physics by emphasizing that laboratory observations should be interpreted against the background of socially constructed theory, rather than used as empirical "proof" that theory is correct. The results of the study indicated that only about 20% of the teachers in the treatment group changed their fundamental beliefs about the purpose of lab work in physics. However, most of the teachers indicated that they paid more attention to the goals of their labs and took more care planning their labs than before the intervention. The Lavonen et al. study (2004) indicates that experienced teacher belief change is difficult, even with well-planned and extensive inservice education.

A study by Trumbell, Scarano and Bonney (2006) illustrates how teachers' core beliefs can influence both classroom practice and belief change. The researchers investigated the teaching and nature of science beliefs for two veteran middle school teachers involved in professional development program to support the implementation of inquiry in the classroom. Both teachers carried out inquiry-based projects with their students and both struggled with how to structure these in their life science classes. One participant, Natalie espoused beliefs about the nature of science that were largely in line with reform-based views, however, she enacted a superficial version of inquiry. Her beliefs in the importance of structure, clarity and the direct transmission of knowledge limited her willingness to let students ask questions or be confused. One core element of her belief structure appeared to be her reluctance to let her students make any mistakes; "she seemed to lack faith in her students' ability to learn" (Trumbell et al., 2006, p. 1741). Natalie's beliefs about science teaching and learning remained static over the three years of the study.

In contrast, the second participant, Meryl, was willing to accept uncertainty and "messiness" in science instruction. She continued to try various inquiry approaches over the course of the study. Meryl approached teaching as research, consistently trying out innovations, gathering feedback and adjusting instruction. The willingness to learn from instructional experimentation may be of central importance for the possibility of teacher belief change. Meryl's teaching practices and beliefs about the nature of science gradually changed in a parallel fashion. Her inquiry-based teaching evolved over the course of the study until both she and her students were confident in doing fairly sophisticated projects. Meryl's beliefs about the nature of science also changed towards an understanding that science is a thinking process more than a discovery of facts.

Waters-Adams (2006) also investigated the relationship between nature of science understandings and science teaching beliefs for four primary grades (children aged 4-6) teachers in England in a year-long qualitative study. The teachers were concomitantly engaged with conducting action research on their own practice. All four of the teachers initially described science as having a hypothetico-deductive epistemology, in which scientific process or problem-solving skills are used to generate and test hypotheses. However, at the beginning of the study three of the teachers' science instruction was conceptually unfocused and characterized by the transmission of facts. Thus, for three of the teachers there was a disconnect between what they thought science was like and how they taught science with none of the teachers exhibiting inquiry-based science. At the same time, all four teachers held strong beliefs that young children should learn through active engagement, inquiry, exploration and induction. They believed that the teacher should take on the role of facilitator in these endeavors. Through their action research and reflection, the teachers explored how their teaching practices articulated with their views of science. As they began to teach science utilizing their core beliefs about young children, rather than their nature of science views, their instruction became more exploratory and inductive in nature.

To summarize, several researchers have been interested in promoting more accurate views of science as inquiry among experienced teachers and connecting these ideas to classroom actions. Although teachers' core beliefs are resistant to change, two routes to belief and practice change have emerged from this research. First, making teachers aware of core beliefs they already hold that may support reform (e.g. a belief in child-centered instruction or that children can learn from their mistakes) may foster changes in practice that align with exploration and inquiry. Second, the research indicates that teacher action research and reflection over a long period of time holds promise for fostering more deeply rooted change.

PRESERVICE OR NOVICE TEACHER BELIEFS AND THEIR INFLUENCE ON TEACHING

Research suggests that beginning teachers' practices are often related to their needs to keep students managed and engaged, regardless of their beliefs about the most effective forms of science instruction. Talanquer and colleagues (Talanquer, Novodvorsky & Tomanek, 2010) found that beginning science teachers in their Southwestern U. S. study selected activities that were almost always driven by one of the following goals: (a) motivating students; (b) developing science process skills; or (c) engaging students in structured science activities. The authors posited that the early adoption of these goals can lead to the construction of a belief set that prioritizes minimizing disruption over conceptual learning. The findings of the Talanquer study echo those of a previous study by Enochs and colleagues (Enochs, Scharmann & Riggs, 1995) in which preservice elementary teachers self-efficacy beliefs about teaching science were significantly related to their beliefs about pupil control, in addition to their background science preparation and self-perception of effective science teaching.

Instability of Novice Science Teachers' Beliefs

In contrast to experienced teachers, research on preservice or novice science teachers often indicates that their belief systems are disconnected, not well developed and unstable. Novice teachers may hold many competing belief sets which change or "wobble" (Simmons, Emory, Carter, Coker, Finnegan & Crockett, 1999) frequently. Simmons and her colleagues studied the beliefs of 116 beginning science teachers who had recently graduated from 10 different universities in the U. S. They found that the vast majority of these beginning teachers' beliefs "wobbled" between more teacher-centered and more student-centered beliefs about what students should be doing in the classroom. Although the new teachers viewed their own teaching as decidedly student-centered, their actual teaching practices were predominantly teacher-centered.

Yilman-Tuzman and Topcu (2008) investigated the inter-relationships among epistemological beliefs, epistemological world views and self-efficacy beliefs of 429 Turkish science preservice teachers. They found that the teachers' epistemological beliefs were not well-developed and that their survey scores for different aspects of epistemology varied widely. For example, the participants held sophisticated beliefs about the epistemological dimension "innate ability." That is, the teachers largely exhibited the view that children's intelligence is not fixed and can be developed through good teaching practices. However, their epistemological understandings of "simple knowledge," whether there can be more than one right answer, and "certain knowledge," whether knowledge is fixed, were much less sophisticated. The participants held positive beliefs about teaching with student-centered strategies, but also voiced their strong preference for having students memorize facts. The preference for memorization was found in all of the teachers, regardless of their self-efficacy beliefs.

Similarly, Luft and her colleagues (Luft, Firestone, Wong, Ortega, Adams & Bang, 2011) found that science teachers within the first three years of service held unstable beliefs about student-centered versus teacher-centered learning. Most of their participants held more teacher-centered views when beginning their first year, although these changed somewhat during the second year towards more student-centered beliefs, especially for those in science specific professional development groups. Interestingly, during the third year, the teachers' beliefs tended to shift back towards a more teacher-centered orientation. However, those who received science specific professional development support continued to implement more student-centered strategies in their practice. The implication was that once these practices were in place, they continued to be used by the teachers despite their shift back towards more teacher-centered beliefs.

Therefore, some research findings suggest that the early years of teaching offer an impressionable period that provides opportunities for change in beliefs and practice. Yilman-Tuzman and Topcu (2008) suggested that learning epistemologies be directly taught in preservice teacher education courses. Luft and colleagues (2011) asserted that science specific professional development during the induction years is a key way to influence practice towards more student-centered orientations. There is also some evidence that the views of mentor teachers influence the beliefs of preservice teachers (Boz & Uzuntiryaki, 2006; Crawford, 2007) with the implication being that mentors be chosen for their positive beliefs about reform-based teaching.

Beliefs and Knowledge about Teaching Evolve Together

Another form of inconsistency between beliefs and practice arises when novice teachers espouse positive, yet peripheral, beliefs about reforms such as inquiry, yet lack understanding of the learning sciences, content knowledge or pedagogical content knowledge to carry these out reforms in the classroom. Boz and Uzuntiryaki (2006) found that most of the preservice chemistry teachers in their Turkish study failed to develop constructivist-oriented beliefs about teaching and learning during practice teaching. Even when teachers in their study espoused positive beliefs about student-centered strategies such as group work, they did not have a deep understanding of how those strategies promoted learning. The fact that most preservice teachers have experienced years of traditional science instruction is often cited as a barrier to forming more reform-based beliefs (Boz & Uzuntiryaki, 2006). Crawford (2007) asserted that preservice teacher beliefs about both science and science teaching are the most powerful influences on whether novice teachers implement inquiry-based instruction in the classroom, although these vary widely for preservice teachers. She suggests that frequent or widespread implementation of inquiry-based teaching may not be a practical expectation for novice science teachers who must learn a repertoire of teaching skills rapidly.

Veal (2004) provided an in-depth qualitative study of various factors influencing the beliefs and practices of two preservice chemistry teachers. First, he found that contexts in which the students learned chemistry greatly influenced their knowledge and beliefs. One participant had learned chemistry in an academic context, having had the opportunity to be an undergraduate research assistant. The other participant learned more about the practical applications of chemistry through her experience working in a veterinary clinic. These background contexts influenced the knowledge, beliefs and ways in which these novice teachers translated chemistry for their students in the classroom.

Further, Veal indicated that the beliefs of the two participants did change over time in concert with the development of their pedagogical content knowledge (PCK). He asserted that, "beliefs informed the practice of the participants in the classroom, and knowledge gained in the classroom informed the participants' beliefs" (Veal, 2004, p. 46). Veal suggested that the two participants' beliefs acted as a filter for the development of PCK, guiding the direction of learning through experience. This complex relationship between beliefs and PCK has important ramifications for teacher education. Veal suggests that teacher candidates could enhance their own PCK by first exploring their own knowledge assumptions, making their epistemologies about science and learning explicit, and then examining teaching applications that match those epistemologies.

In summary, preservice and novice teachers are at risk for adopting belief sets that support classroom management and reinforce practices that keep students busy. New teacher beliefs are unstable or "wobbling" beliefs and become more fixed over the first few years of teaching. This induction period may be a prime opportunity for novice teachers to explore their own epistemologies and beliefs about the nature of science and cultivate PCK that is compatible with their beliefs. Online and on-site science specific professional development activities hold promise for shaping beliefs and practices that support teaching for meaningful learning. Further research on the ways teacher educators can help new teachers unpack their epistemologies is a logical next step in this field.

CONCLUSIONS AND IMPLICATIONS FOR RESEARCH

Taken together, the studies reported on in this overview of science teacher beliefs point to a few synthesized understandings of science teacher beliefs. First, experienced teachers' core beliefs have a strong impact on both their enactment of the curriculum and their stance towards implementing reform-based practices. As well, teachers undergoing professional development may adopt new ideas about learning or the nature of science, but these are often held peripherally and are not easily integrated into their core belief sets. These findings imply that working with experienced teachers' core beliefs is a natural starting place for professional development. It may be useful for science teacher educators to help teachers unpack their core beliefs and reflect on what their own beliefs mean for practice. Little is known about whether an increased self-awareness of one's own core beliefs and their pedagogical implications can affect belief change.

Research shows that those teachers who are responsive to reform-based teaching have underlying philosophical values about children, learning and the role of education that are broad-brushed and positive (Levitt, 2001; Wallace & Priestley, 2011; Waters-Adams, 2006). For example, reform-minded teachers tend to believe that: (a) children are capable of high level thinking; (b) learning how to learn is an important purpose of schooling; (c) promoting thinking is more important than conveying factual knowledge; (d) learning involves making mistakes; (e) curriculum should be largely student-centered; and (f) a teacher's primary role is that of facilitator of learning (Levitt, 2001; Priestley et al. 2011, Trumbell et al., 2006; Wallace & Kang, 2004; Wallace & Priestley, 2011; Waters-Adams, 2006). Making this research available to teachers through professional development activities might cause teachers to question their own core beliefs or reflect on their practices. Evoking cognitive dissonance, for example about students' capability for high-level thinking, may support teacher belief change towards reform-based teaching. Providing research on these novel approaches to professional development is an important research agenda for the field.

Second, there is evidence that both novice and experienced teachers can make lasting changes to their practices even without changing their core beliefs (Lavonen et al., 2004; Levitt, 2001; Luft et al., 2011, Waters-Adams, 2006). How and why this phenomenon has been observed is not entirely clear. It may be that incorporating particular practices into routines and teaching repertoires, even if required for coursework, can lead to their regular use. Students' positive responses to these practices may stimulate teachers to continue their use. Or perhaps, teachers using these reform-based practices are, in fact, in the process of slowly changing their beliefs. Therefore, it may be that adopting a new set of practices can lead to belief change, just as belief change can lead to new practices. This would imply that science teacher educators should focus on teaching practices that are reform-based, but also appealing to teachers for other reasons (e.g. promoting student engagement). Changes to practice may be one entry point in a cycle of belief and practice change. More research is needed to explore the complex relationship between beliefs and practices.

Third, there is a need for more research on the formation of teacher beliefs in early stages of teaching, including the induction phase (Luft et al., 2011). The research cited in this chapter indicates that the first few years of teaching are probably the most critical for the formation of reform-based teaching beliefs. If researchers could pinpoint more precisely the types of experiences and reflection that lead to positive views about reform, these could be replicated more often. In-depth, longitudinal case studies of how novice science teachers build their beliefs over time like that of Veal (2004) may useful, however, there is a complex interaction between belief sets that students have before entering teacher education programs and how teacher education shapes those beliefs (Avraamidou, 2013). The influence of mentor teachers' beliefs on novice teacher beliefs is also an important area of study, although space precludes a discussion of this large topic here.

Finally, there is a concern that even when science teachers hold very positive beliefs about reform-based teaching, they are thwarted from enacting these in the classroom due to educational policy in the current political climate. Perhaps this situation will change with the introduction of the Next Generation Science Standards (Achieve Inc., 2013) into school culture. The implementation of reform-based standards and concomitant changes to science assessment may provide the impetus needed for bridging the research and practice gap in science education. The science education community will undoubtedly be interested in science teachers' beliefs about the new standards and their implementation.

REFERENCES

Achieve, Inc. (2013). *The next generation science standards.* Retrieved August 7, 2013, from http://www. nextgenscience.org/

Avraamidou, L. (2013). Prospective elementary teachers' science teaching orientations and experiences that impacted their development. *International Journal of Science Education, 35*(10), 1698–1724.

Au, W. (2007). High stakes testing and curricular control: A qualitative metasynthesis. *Educational Researcher, 36*(5), 258–267.

Ball, S. J. (2003). The teacher's soul and the terrors of performativity. *Journal of Education Policy, 18*(2), 215–228.

Boz, Y., & Uzuntiryaki, E. (2006). Turkish prospective chemistry teachers' beliefs about chemistry teaching. *International Journal of Science Education, 28*(14), 1647–1667.

Bryan, L. A. (2012). Research on science teacher beliefs. In B. Fraser et al. (Eds.), *Second International Handbook of Science Education* (pp. 477–495). Springer International.

Bryan, L. A. (2003). The nestedness of beliefs: Examining a prospective elementary teacher's beliefs about science teaching and learning. *Journal of Research in Science Teaching, 40,* 835–868.

Cotton, D. R. E. (2006). Implementing curriculum guidance on environmental education: The importance of teacher beliefs. *Journal of Curriculum Studies, 38*(1), 67–83.

Crawford, B. A. (2007). Learning to teach science as inquiry in the rough and tumble of practice. *Journal of Research in Science Teaching, 44,* 613–642.

Cronin-Jones, L. (1991). Science teacher beliefs and their influence on curriculum implementation: Two case studies. *Journal of Research in Science Teaching, 28,* 235–250.

Dictionary.com. (2012) Retrieved July 25, 2012, from http://dictionary.reference.com/browse/beliefs?s=t

Enochs, L. G., Scharmann, L. C., & Riggs, I. M. (1995). The relationship of pupil control to elementary science teacher self-efficacy and outcome expectancy. *Science Education, 79*(1), 63–75.

Green, T. F. (1971). *The activities of teaching.* Tokyo: McGraw-Hill.

Haney, J. J., Czerniak, C. M., & Lumpe, A. T. (1996). Teacher beliefs and intentions regarding the implementation of science education reform strands. *Journal of Research in Science Teaching, 33,* 971–993.

Hursh, D. (2007). Assessing no child left behind and the rise of neoliberal education policies. *American Educational Research Journal, 44*(3), 493–518.

Kang, N. H., & Wallace, C. S. (2004). Secondary science teachers' use of laboratory activities: Linking epistemological beliefs, goals and practices. *Journal of Research in Science Teaching, 89,* 140–165.

Kelly, G. A. (1955). *The psychology of personal constructs.* New York, NY: Norton.

Levitt, K. (2001) An analysis of elementary teachers' beliefs regarding the teaching and learning of science. *Science Education, 86,* 1–22.

Lingard, B. (2005). Socially just pedagogies in changing times. *International Studies in Sociology of Education, 15*(2), 165–187.

Lavonen, J., Jauhiainen, J., Koponen, I. T. & Kurki-Suonio, K. (2004). Effect of a long-term in-service training program on teachers' beliefs about the role of experiments in physics education. *International Journal of Science Education, 26*(3), 309–328.

Luft, J. A., Firestone, J. B., Wong, S. S., Ortega, I., Adams, K., & Bang, E. J. (2011). Beginning secondary science teacher induction: A two-year, mixed methods study. *Journal of Research in Science Teaching, 48*(10), 1199–1244.

Lumpe, A., Haney, J., & Czerniak, C. (2000). Assessing teachers' beliefs about their science teaching context. *Journal of Research in Science Teaching, 37,* 275–292.

Munby, H. (1984). A qualitative approach to the study of a teacher's beliefs. *Journal of Research in Science Teaching, 21*(1), 27–38.

Nespor, J. (1987). The role of teacher beliefs in the practice of teaching. *Journal of Curriculum Studies, 19,* 317–328.

Olson, J. (1981). Teacher influence in the classroom: A context for understanding curriculum translation. *Instructional Science, 10,* 259–275.

Pajares, F. (1992). Teachers' beliefs and educational research: Cleaning up a messy construct. *Review of Educational Research, 62,* 307–332.

Priestley, M., Edwards, R., Priestley, A., & Miller, K. (2012). Teacher agency in curriculum making: agents of change and spaces for manoeuvre. *Curriculum Inquiry, 42*(2), 191–214.

Ranson, S. (2003). Public accountability in the age of neo-liberal governance. *Journal of Educational Policy, 18,* 459–480.

Richardson, V. (1996).The role of attitudes and beliefs in learning to teach. In J. Sikula (Ed.), *The Handbook of Research in Teacher Education* (pp. 102–109*).* New York, NY: Macmillan.

Rop, C. J. (2002). The meaning of students' inquiry questions: A teacher's beliefs and responses. *International Journal of Science Education, 24*(7), 717–736.

Simmons, P., Emory, A., Carter, T., Coker, T., Finnegan, B., & Crockett, D. (1999). Beginning teachers: Beliefs and classroom actions. *Journal of Research in Science Teaching, 36,* 930–954.

Talanquer, V., Novodvorsky, I. & Tomanek, D. (2010). Factors influencing entering teacher candidates' preferences for instructional activities: A glimpse into their orientations towards teaching. *International Journal of Science Education, 32*(10), 1389–1406.

Trumbell, D. J., Scarano, G., & Bonney, R. (2006). Relations among two teachers' practices and beliefs, conceptualizations of the nature of science, and their implementation of student independent inquiry projects. *International Journal of Science Education, 28*(14), 1717–1750.

van Driel, J. H., Beijard, D., & Verloop, N. (2001). Professional development and reform in science education: The role of teachers' practical knowledge. *Journal of Research in Science Teaching, 38,* 137–158.

van Driel, J. H., Bulte, A. M. W., & Verloop, N. (2005). The conceptions of chemistry teachers about teaching and learning in the context of a curriculum innovation. *International Journal of Science Education, 23*(3), 303–322.

Veal, W. R. (2004). Beliefs and knowledge in chemistry teacher development. *International Journal of Science Education, 26*(3), 329–351.

Wallace, C. S., & Kang, N.-H. (2004). An investigation of experienced secondary science teachers' beliefs about inquiry: An examination of competing belief sets. *Journal of Research in Science Teaching, 41,* 936–960.

Wallace, C. S., & Priestley, M. (2011). Teacher beliefs and the mediation of curriculum in Scotland: A socio-cultural perspective on professional development and change. *Journal of Curriculum Studies, 43*(3), 357–381.

Wallace, C. S. (2013). Policy and the planned curriculum: Teaching high school biology every day. In M. Dias, C. Eick & L. Brantley-Dias. (Eds), *Science teacher educators as K-12 teachers: Practicing what we teach.* Springer International.

Waters-Adams, S. (2006). The relationship between understandings of the nature of science and practice: The influence of teachers' beliefs about education, teaching and learning. *International Journal of Science Education, 28*(8), 919–944.

Yerrick, R., Parke, H., & Nugent, J. (1997). Struggling to promote deep rooted change: The filtering effect of teachers' beliefs on understanding transformational views of teaching science. *Science Education, 81,* 137–159.

Yilmaz-Tuzan, O., & Topcu, M. S. (2008). Relationships among preservice science teachers' epistemological beliefs, epistemological world views and self-efficacy beliefs. *International Journal of Science Education, 30*(1), 65–85.

AFFILIATION

Carolyn S. Wallace
Center for Science Education
College of Arts and Sciences
Indiana State University

SECTION TWO

ORIENTATIONS TOWARDS BELIEFS RESEARCH

ROBERT H. EVANS

CULTURAL EFFECTS ON SELF-EFFICACY BELIEFS

INTRODUCTION

It is tempting to use student achievement scores to understand teaching and pupil learning. However, looking at other attributes between pupils and teachers in science classes within schools and between schools and regions, and in even larger political areas within countries, such as provinces and states, holds the promise of valuable insight. For instance, large international studies of pupils such as Trends in International Mathematics and Science Study (TIMSS) (Provasnik, Kastberg, Ferraro, Lemanski, Roey, & Jenkins, 2012) and the Programme of International Student Assessment (PISA) (Organisation for Economic Co-operation and Development (OECD), 2010) found that while cross-national comparisons of achievement were informative, they could not stand alone. Fortunately, the researchers associated with these studies also collected information relevant to pupil attributes, school and home environments, curricular differences and other learning factors that provided insights into student learning. Such ancillary information was necessary to contextually interpret and compare cross-national achievement scores (OECD, 2010).

Context was also important to the Teaching Practices and Pedagogical Innovation Study (TALIS). Researchers in this study showed significant differences in teacher's self-efficacy across participating countries, and that the classroom practices of teachers were strongly influenced by traditions, culture and educational policies (Vieluf, Kaplan, Klieme, & Bayer, 2012). With closer examination of different variables associated with context, researchers concluded that when teachers engaged in different teaching methods and were active in learning communities, they received more feedback about their teaching, participated in more professional development activities, and as a result reported higher self-efficacies. They also found that frequent use of multiple teaching methods were correlated with constructivist beliefs and higher self-efficacies.

WITHIN CULTURE COMPARISONS OF PUPIL SELF-EFFICACY BELIEFS

'Culture is mental programming' that is acquired early and expressed throughout life (Hofstede, Hofstede, & Minkov, 2010). It is also known as shared thinking, feeling and includes actions, all of which are learned from home, school and communities (Hofstede et al., 2010). 'Nationalities' are one cultural classification in which a group shares perceptions of meaning related to daily encounters (Geertz, 1994).

R. Evans et al., (Eds.), The Role of Science Teachers' Beliefs in International Classrooms, 35–48.

However, since there is cultural variance within nationalities, studies of self-efficacy between and within countries, municipalities, and state boundaries require additional contextual information to allow for valid comparisons. Using measures of self-efficacy to make within-cultural comparisons of pupils and teachers may require even more caution than when comparing achievement scores. This is because self-efficacy is not as dependent upon variations in curricular goals and curricular implementation. Instead, the creation of self-efficacy is highly dependent on the cultural surroundings.

The notion of nationalities also extends to classrooms. Bandura (1997) suggested four sources that influence personal capacity beliefs: *mastery experiences, vicarious experiences, verbal persuasion and physiological/affective differences*. He further added that these sources are affected by small and large cultural differences. For example, a science teacher's individual style of pupil feedback can affect the pupils in that classroom and not in the classroom next door. Both teacher and pupil self-efficacies are influenced by the micro-culture in which they operate. With this added dimension, teacher and pupil self-efficacy scores may not be simply compared to one another. Instead, researchers must acknowledge that there will be cultural differences from science classrooms to extreme geographic regions of a single country.

One example of such sensitivity to local change of self-efficacy is reported by Marx, Ko and Friedman (2009). They concluded that after the election of Barack Obama in the United States in the fall of 2008, the test-taking success of African-Americans increased significantly. They attributed the improved performance to changes in anxieties about racial stereotypes among the African-American students. One potential explanation for the increased self-efficacy of African-American students was the effect of a *vicarious experience*. Barack Obama became a role model for African-American students. In this case, the influence of the 'vicarious experience' was limited by racial boundaries, but well within the change modalities suggested by Bandura (1997).

Since factors that influence self-efficacy such as *mastery experiences, vicarious experiences, verbal persuasion* and *physiological/affective differences* (Bandura, 1997) vary according to the socio-cultural environments in which they are experienced, the results from instruments which measure self-efficacy are likely to also vary. For pupils, this means that the large variability within, for example, home and school environments is likely to affect these four sources of efficacy judgment. For example, a group of teachers at a school, who provide exceptional and credible amounts of *verbal persuasion* to their pupils about their abilities to perform science experiments, are likely to increase pupil self-efficacy scores on a given instrument over a similar school that does not have such vigorous coaching. Of course, this does not necessarily invalidate the efficacy scores' ability to predict effort, but the chance exists that verbally persuaded students may have higher self-efficacy scores than pupils in a less-coached environment. While this could be a desirable school outcome, it may not be possible to compare self-efficacy scores between students from different environments without adequately describing their contexts.

Teacher's efficacies may also be influenced by different school contexts. Science teachers who work with strong principals may have higher self-efficacies. This may be a result of various outcomes associated with the principal, such as more supportive verbal persuasion, better laboratory equipment and facilities and a culture of motivation and rewarded effort. Communities of teachers may also create higher efficacies, as the vicarious attributes of peer-to-peer modeling and coaching have the potential to influence how teachers perceive their efforts and success.

These contextual differences, and many others, are prevalent in every type of learning environment for both teachers and pupils. Consequently, instruments designed to measure such self-efficacies can be improved by including contextual questions. The information gleaned from these contextual questions can then be used to understand the study results.

BETWEEN CULTURAL COMPARISONS OF PUPIL SELF-EFFICACY BELIEFS

Studies of self-efficacy that cross international boundaries reveal the influence of learning environments on pupils. A quasi-experimental study conducted by Little, Oettingen, Stetsenko and Baltes (1995), looked at the differences in formative self-efficacy influences in schools in three different cultures and reported significant differences in pupil self-efficacy. They specifically found that German and Russian pupils had significantly lower self-efficacies than did comparable American pupils. They also found that the American students' self-perceptions correlated to a lesser degree with teacher assigned grades than did those of the German and Russian students. They hypothesized that the higher American self-efficacies could be due to three factors, 'degree of dimensionality of the school', 'feedback directness' and 'feedback transparency'.

Applying the first factor (degree of dimensionality), Oettingen (1995) determined that the German and Russian curricula were fairly uniform and basically the same for all students. This means that at a given curriculum level, all pupils were given the same materials, assignments and goals without taking into account their individual differences (Little, Lopez, Oettingen, & Baltes, 2001). By contrast, the American school curricula were characterized as multidimensional. Specifically, American schools had more cooperative and individualized learning opportunities with individual learning needs more frequently guiding instruction.

In this example, dimensionality may have affected the self-efficacy of the pupils. Specifically, it is simply harder for pupils to experience '*self-mastery*' in a multidimensional curriculum such as in the American curricula, which emphasizes cooperative work. For American pupils, there are fewer students engaged in the exact same tasks at the same time, and as a result there are fewer pupils with whom to compare one's performance. *Self-mastery experiences* are a part of Bandura's (1997) group of influences on self-efficacy. For individuals, these experiences can elevate or deflate their personal self-efficacies by comparing their success of a performance to the performance of others. Aside from the multidimensional curriculum, American teachers created learning environments that provided opportunities for changes in self-efficacy. The teachers worked to provide tasks to individual students at which they were likely to succeed; therefore, potentially enhancing individual self-efficacy. They were observed praising student efforts which may have boosted pupil self-efficacies through *verbal persuasion*. In the small cooperative American groups, student self-appraisals were continuously modified by *verbal feedback* from other students and by *vicarious* emulation to group members. On balance, these factors seemed to have a greater 'raising' impact on student self-efficacies than self-mastery had on 'lowering' them.

On the other hand, pupils in the more unidimensional curricular environments of the European countries had many more daily opportunities to compare their overall success on comparable tasks with their peers. They did not have the amelioration and individualization of small cooperative groups to sway their *self-mastery* judgments. This more competitive feedback without as much individual praise and consequently less *verbal persuasion*, may have contributed to the significantly lower efficacy beliefs of the German and Russian pupils.

The second relevant factor, 'feedback directness' also differed between the European and American samples. Little et al. (2001) reported that the daily feedback in the American sample was designed to raise student's performance expectations. These teachers often had comments that praised student partial successes and effort, while the feedback of the German and Russian teachers was more critical and often consisted of statements of correctness. The reduced critical orientation of the feedback provided by American teachers was aimed at building confidence and raising pupil self-efficacies, and provided a form of *verbal persuasion*. Conversely, the consistently more critical feedback found in the European settings, would have the effect of either maintaining or lowering pupil self-efficacies.

Thirdly, the 'transparency of the feedback', the degree to which it was either public or private in different cultures, may have contributed to pupil self-efficacies by affecting self-perceptions of *mastery experiences*. In a mostly private feedback environment (e.g., written comments, grades, reports to parents and individual comments), pupils have fewer opportunities to accurately assess their self-efficacies at tasks since they don't know how well others are doing compared to themselves. These relatively lower levels of comparative self-reflection, accompanied by esteem building teacher feedback, may have elevated self-efficacy levels in the American sample. By contrast, in mostly public and realistic feedback situations (e.g., critical feedback in front of an entire class), as experienced particularly in the former East Germany, pupils were more likely to judge their success or mastery of tasks either more realistically or even somewhat more harshly, consequently lowering their self-efficacy levels.

These three factors are particularly relevant to science classrooms. For example, when students are conducting classroom experiments there are frequent opportunities for self-efficacy judgments. When a class of students replicates a 'cook-book' or verification experiment, there is a reduction in the opportunities for multidimensional individual and small group independent work. However, in classrooms with more independent collaborative group work, where teacher feedback is aimed at encouraging individual students, there is an increased chance for the growth of student self-efficacy. The implications for science teachers are important. When science teachers adopt methods of inquiry instruction there are more opportunities for increased student success and self-efficacy building feedback. It would be useful for future research to examine the potential self-efficacy outcomes during the use of inquiry methods in science classes.

Even more relevant for comparisons of self-efficacy are the results of a follow-up study with German and American schools after the reunification of Germany. Little et al. (2001) found that even though the structural and organizational aspects of the East German schools were modified to coincide with those of West Germany, which were somewhat closer to the American sample in forms of feedback and curriculum, the self-efficacies of the East German children did not change significantly. They hypothesized that either the self-assessments of efficacy were highly stable or that the East German teachers maintained previous methods of feedback and curriculum use, despite the overall organizational changes (Little et al., 2001). However, the correlation between East German pupil's efficacies and their grades decreased significantly when compared to the level of West German pupils. This could be due to changes in performance feedback, which were no longer as harsh and critical or it could be the result of changed attitudes towards grades.

Teacher self-efficacy beliefs are also influenced by the same factors as pupils. For example, in schools, states and countries where the feedback teachers receive about their efforts persuades them that they are performing well, their self-efficacies will rise. However, when they have either infrequent feedback from peers due to simultaneous teaching schedules or from only the most successful peers, their self-efficacies may fall.

The contextual differences between the German, Russian and American teaching-learning environments, with regards to self-efficacy variables, are clearly associated with different efficacy assessments. Even within similar cultures, such as those of East and West Berlin, changes in factors known to affect control beliefs can have a variety of affects on self-perceptions of teachers or pupils. In these examples and others, knowing something about the study context is a necessary co-requisite for within culture comparisons of self-efficacy (Little et al., 2001).

RELEVANCE TO WITHIN CULTURAL COMPARISONS

Early on, Bandura (1982) characterized how self-efficacy is best assessed among individuals including teachers and students. In an experiment to find out whether asking individuals to declare their self-efficacy judgments increases the congruence of their performance with these predictions, Bandura and his colleagues found no increase (Bandura, 1982). People don't attempt to rise to the level of action similar to the self-efficacy they have reported to a researcher. On the contrary, there was a decline in the congruence between reports and actions when subjects were publicly asked to assess their self-efficacy. Individuals tended to be modest in their self-appraisals compared to when their answers were totally confidential.

An implication for assessing self-efficacy is that it should be done confidentially so that individuals do not limit their self-assessments due to the potential of social inspection of their assertions. Consequently, the most valid reports of self-efficacy are made in confidence and with personal anonymity. Confidential questionnaires accommodate this human characteristic and personal interviews of any sort contradict it. Hence, the history of self-efficacy assessment is mainly with paper and pencil questionnaires.

An early instrument that measured general teaching self-efficacy and assured participant anonymity was developed in 1984 (Gibson & Dembo). This *Teacher Efficacy Scale* (Gibson & Dembo, 1984) was the basis for the development of later science specific scales: *Science Teaching Efficacy Beliefs Instrument* (STEBI)-A (Riggs & Enochs, 1990), subsequent STEBI-B (Enochs & Riggs, 1990) and various adaptations of these instruments. It also served as a model for the confidential assessment of self-efficacy through the use of paper and pencil questionnaires. However, not all self-efficacy studies report whether or not participants were assured anonymity. This has implications for the collected data, as non-confidential personal estimates can lead to an under-estimation of predicted ability.

Confidentiality in other studies, for various reasons, has not always been maintained. For example, Weinburgh (2007) conducted a qualitative study that depended on the coordinated use of open-ended questionnaires, journals and video recordings of classroom proceedings to assess changes in teacher self-efficacy. Before engaging in an extended experience with 'meal worms,' preservice elementary teachers were asked to estimate to the instructors how they felt about their abilities to teach science and specifically about life cycles. The initially low-self-efficacies of the teachers may have been even lower than concluded, as the teachers may have felt constrained to publically state their ability to work with 'meal worms'. Although there may have been improvement in teacher self-efficacy after the intervention, the actual degree of change may have also been compromised by the initial non confidential reporting of self-efficacy. Some of these pre-service elementary education teachers may have publicly responded and reacted in stereotypical or modest ways to the possibility of teaching with 'meal worms', while privately they were not so skeptical. Of course, such underestimations of predicted ability have the potential to misinform teacher educators and researchers about the effects of their work.

Consequently, when measures of pupil and teacher self-efficacy are formed, even within individual countries, states, communities and schools, they need to consider the corresponding environments. Seeking sources of influence on self-efficacy from each pupil may help determine the relative comparability of scores on a given instrument. Assessing the environment of each teacher and pupil for evidences of Bandura's (1997) four sources of self-efficacy through observation and interview, would also allow for some interpretative comparison of self-efficacy scores within country borders.

Of course, such interviews and observations would not capture all of the school and outside influences on self-efficacy. To capture a portion of these influences, items could be added to self-efficacy questionnaires for pupils and teachers to indicate their exposure to 'environmental influences on self-efficacy'. Examples for pupils and teachers could be as straight forward as illustrated via examples in Table 1.

Table 1. Items which could be added to Self-efficacy Questionnaires for Pupils and Teachers to Indicate their Exposure to 'Environmental Influences on Self-efficacy'

Pupil Mastery Experiences
On your latest science test or report, how did you do compared to the rest of the class?

Not so good									Very good
10	9	8	7	6	5	4	3	2	1

Pupil Vicarious Experiences
How good are your best friends at science?

Not so good									Very good
10	9	8	7	6	5	4	3	2	1

Pupil Verbal Persuasion
How often have people (friends, parents, and or teachers) recently told you that you are pretty good at science?

A lot Never									
10	9	8	7	6	5	4	3	2	1

Pupil Physiological/Affective Differences
When you come into science class, how do you usually feel inside about being there?

Not very good									Very Good
10	9	8	7	6	5	4	3	2	1

Teacher Mastery Experiences
Looking back at all of the science lessons you taught last week, in general, how were they?

Very good									Not so good
10	9	8	7	6	5	4	3	2	1

Teacher Vicarious Experiences
How good is the teaching of your science teaching colleagues?

Very good									Not so good
10	9	8	7	6	5	4	3	2	1

Teacher Verbal Persuasion
During the past year, how often have you been praised as a science teacher by students, parents, colleagues and/or administrators?

A lot									Never
10	9	8	7	6	5	4	3	2	1

Teacher Physiological/Affective Differences
When you come into your science class, how do you usually feel inside about being there?

Not very good									Very Good
10	9	8	7	6	5	4	3	2	1

IMPLICATIONS FOR BETWEEN CULTURE COMPARISONS
OF SELF-EFFICACY BELIEFS

The implications for between culture comparisons of self-efficacy are the same as those for within cultures. Because of the variability of sources of self-efficacy between all contexts both within and between cultures, comparisons of self-efficacy without substantial contextual information are of little value. Changes in teacher and student self-efficacies due to variations in teaching and learning can be followed within specific learning communities when there is a description of the contextual elements. The use of extra questionnaire items such as those suggested in Table 1 or focused interviews to identify the contextual influences on self-efficacy are necessary to fully understand changes in self-efficacies of teachers and students.

Comparisons of self-efficacy between cultures also require attention to contextual and societal variables. This can be addressed in the design of the instruments. For researchers, it is important to attend to the variation between and within cultures. This is because anthropological culture is learned and the environments where such learning occurs varies widely. As a result, personal expression, which includes self-efficacy, is to some extent a result of the culture.

Hofstede et al. (2010) used a correlational analysis to look at the national cultural values of people in over seventy countries. They drew their data from a series of large international surveys conducted from the 1970s to 2002. From their analysis, Hofstede et al. (2010, p. 30) defined four areas of cultural difference:

Social inequality, including the relationship with authority

The relationship between the individual and the group

Concepts of masculinity and femininity: the social and emotional implications of having been born as a boy or as a girl

Ways of dealing with uncertainty and ambiguity, which turned out to be related to the control of aggression and the expression of emotions

Each of these areas of cultural diversity can have differential effects on the perceptions of self-efficacy among teachers and pupils by shaping their responses to challenges. These areas also affect self-reported measures of efficacy because they influence an individual's relationship to authority and to the greater society, and they contour perceptions of gender roles and the control of aggression and feelings (Hofstede et al., 2010).

Social Inequality

Since substantial learning of 'culture' occurs during schooling, teachers and students affect individual self-perceptions. The concept of 'power distance' is particularly relevant to an understanding of how social inequalities influence teaching and learning, and concomitantly self-efficacies. 'Power-distance' is explained as the degree to which less powerful people in society, including pupils in schools, expect and accept that there is an unequal distribution of power in their world (Hofstede et al., 2010). In 'big power distance' countries, pupils expect teachers to create a large emotional distance between themselves and their students. As a result, teachers are treated deferentially, the instruction is teacher-centered and students speak only when recognized by the teacher. Because of the power and respect given to teachers, they are often not criticized by students in public. Consequently, teachers often influence pupil self-efficacy beliefs through various forms of performance feedback on *mastery experiences* and through *verbal persuasion*. However, teachers who have a large emotional distance between themselves and their students rarely receive feedback from those students, and this reduces the potential effects of student feedback on teacher self-efficacy.

In 'low power distance' countries, teachers treat students more as equals. In countries where there is a low power distance between teachers and students, there is a student-centered environment and spontaneous interactions between students and teachers. Hence, many opportunities exist for feedback and self-assessment between teachers and students. In this more student-centered environment, self-efficacies for both students and teachers vary according to the frequency of feedback. Examples of countries or regions with a big power-distance include Malaysia, Slovakia, Guatemala, Panama, the Philippines and Russia, while some countries with a low power distance include New Zealand, Denmark, Israel and Austria (Hofstede et al., 2010).

One possible outcome of the difference in power relationships is the lack of attainment of independence by less-talented pupils in low-power distance situations. More able pupils in low-power distance schools are able to achieve independence, but this is not the case for their less-able peers (Hofstede et al., 2010). Consequently, the higher self-efficacies which come from self-appraisals of mastery experiences in independent situations will not be as common among the less-talented pupils. This may also be true of lower-class students who more typically come from high-power differential homes (Hofstede et al., 2010). When schooled in low-power classrooms, they may fail through lack of acculturation to raise their personal self-efficacies from independent mastery experiences.

Relations Between the Individual and the Group

Also relevant to teacher and pupil development of self-efficacy in different cultures is the underlying relationship between individuals and larger groups within each culture. In invidualist societies where the connections between individuals are not tight, each member more or less attends to themselves and their family. However, in more collectivist cultures, citizens are a part of strong cohesive groups which care for each other throughout their lives. Examples of where individualism is strong include the United States, Australia and Great Britain, whereas countries with stronger collectivism are Spain, Japan and South Korea (Hofstede et al., 2010).

When the development of self-efficacy is considered, the opportunity for *mastery experiences* may be constrained in collectivist cultures where there is encouragement for whole group achievement rather than individual accomplishment. In addition, the

43

verbal persuasion common in cultures that nurture and reward the achievements of individual teachers and students may be more directed at whole group achievement in more collective cultures. Consequently, self-efficacies will develop in different ways between individualist and collectivist cultures.

In the classroom, one example of behavioral difference across cultures is in students' answers to questions posed by a teacher to the entire class. In classrooms in collectivist cultures, pupils tend to not volunteer to answer teacher questions since doing so without group consultation seems inappropriate. Rather than individually asserting one's knowledge to the group at large, collectivist students prefer to discuss teacher questions in small groups and then appoint a member to share the group consensus with the class (Hofstede et al., 2010). When considering self-efficacy, in collectivist communities *verbal persuasion* may be enhanced by small group discussion and the reinforcement of individual group member ideas by fellow students. The subsequent teacher assessment may also reinforce the contributing group member's answer. On the other hand, where individuals are licensed by their cultures to independently answer teacher questions, their personal *mastery experience* enhances their self-efficacy.

From the teacher's perspective, the individual pupil responses from questions they ask the class may reduce their personal self-efficacy as teachers, and it may limit the number of questions they ask due to negative experiences with questioning. Ultimately, the ability of a teacher to give self-efficacy inducing feedback to individual pupils may be diminished due to the collective behavior of classes.

Fortunately, science teaching and learning which emphasizes inquiry instruction, works well with both individualist and collectivist communities. Inquiry instruction provides many opportunities for enhanced teacher and student feedback due to small group work and frequent individual feedback from both peers and teachers. These private, small group communications influence self-efficacy among teachers and students.

Concepts of Masculinity and Femininity

The typical roles each culture associates with masculinity and femininity can also result in variations in the development of self-efficacy among pupils and teachers. Hofstede et al. (2010) define masculine societies as those where there tend to be distinct differences in the emotional gender roles of men and women. In such cultures, men are expected to be " . . . assertive, tough and focused on material success, whereas women are supposed to be more modest, tender and concerned with the quality of life" (Hofstede et al., 2010, p. 140). Examples of highly 'feminine' countries include Sweden, Norway, Holland and Denmark whereas 'masculine' countries include Japan, Austria, Venezuela and Italy (Hofstede et al., 2010).

In 'feminine' cultures, one goal is to provide equal access to all aspects of the culture to both females and males. The implications of these orientations are particularly strong in the effect of *vicarious experiences* on self-efficacy. In more 'masculine' cultures, students and teachers will more often identify with members of their same gender, thus providing differential sources of personal capacity beliefs for boys and girls, whereas in 'feminine' cultures, such sources are more likely to be cross-gender.

One implication for science pupils is the extent to which they strive to achieve high grades on exams and in coursework. In more 'masculine' societies, such achievement motivation is common among men and women whereas in more 'feminine' cultures, just 'passing' tests or courses is more often the accepted norm. The consequences of these different motivational structures are found in both achievement and in the development of self-efficacy. For example, if students respond to given academic challenges with a need to accomplish them only at a satisfactory level rather than at higher levels, then the definition of *mastery experiences* is changed. More students in 'feminine' cultures may feel they have mastered given tasks since positive feedback will flow to many more individuals and consequently, create higher self-efficacies. Conversely, if only a few students feel they have mastered a given task then self-efficacies due to successful mastery experiences will be diminished, with implications for both learners and teachers. The implications of different efficacies resulting from masculinity and femininity orientations may also include the attractiveness of careers in science to pupils in different cultures.

Ways of Dealing with Uncertainty

How threatened individuals feel by the anticipation of unclear or unknown future situations and how they deal with uncertainty varies substantially from culture to culture (Hofstede et al., 2010). Uncertainly can result in individuals experiencing stress, and it can result in individuals needing rules and norms of behavior that can increase the predictability of events (Hofstede et al., 2010). Countries with high uncertainty avoidance include Greece, Portugal and Japan while those with low profiles are Singapore, Denmark, Sweden and Hong Kong (Hofstede et al., 2010). High stress and anxiety often accompany strong uncertainly avoidance. Consequently, *physiological* reactions are likely to vary profoundly in situations where individuals are attempting new tasks. Such an internal intensification of feelings when trying something 'new' can have both positive and negative effects in terms of changes in self-efficacy. If a teacher in a relatively high uncertainty avoiding culture (e.g., Germany) is convinced to try a new teaching method, whether the outcomes raise or lower the teacher's self-efficacy for using that method will be influenced by stress hormones. Internal feedback to the teacher may overwhelm external indicators of success, such as *verbal persuasion* by others.

In schooling, students from high uncertainly avoiding cultures expect their teachers to 'know it all.' Those teachers who fall short of this goal not only lose the respect of their students, but through such student feedback, suffer from lowered self-efficacies for lessons when their knowledge seems tentative. By contrast, in cultures where uncertainty avoidance is low, teacher's self-efficacies grow through the respect that comes from creating productive academic discussions from what they do not claim to fully 'know'.

Since all of these cultural differences can affect teacher and student self-efficacies through the mechanisms of change defined by Bandura (1997), instruments designed to assess self-efficacy in one culture must be carefully monitored in cross-national comparisons. Without concomitant consideration of the various societal factors affecting self-efficacy development, simple comparisons may not be valid. For example, the results of introducing an inquiry-based science teaching method to different countries and then using changes in teacher and/or student self-efficacies to partially assess the success of inquiry learning on science process skill use, would not be valid without considering the manner in which self-efficacies are moderated in each culture.

CRITICISM OF HOFSTEDE'S NATIONAL CULTURE RESEARCH

McSweeney (2002) raised strong methodological and conceptual concerns with Hofstede's first edition (1991) of 'Cultures and Organizations: Software of the Mind.' These concerns remain relevant to cross-national comparisons of self-efficacy. Principally, McSweeney questioned the claim that populations in one nation share a unique culture. He pointed out that it is invalid to even attempt to characterize national cultures with the idiosyncratic samples used by Hofstede (1991). In Hofstede et al. (2010) this criticism was partially addressed with the addition of a variety of samples gathered since the original analyses.

Such criticism can also be made about some research studies which have looked at self-efficacy across international borders. For example, the reports discussed earlier in this chapter by Little et al. (1995) and Little et al. (2001) consider self-efficacy development in Germany, the United States and Russia. The insights from these studies into the development of self-efficacy in these cultures were useful and the authors were careful to not compare actual self-efficacy scores between countries. However, by using a national name such as The United States to identify measurements in only one small population, the authors of these studies imply a national culture, which is criticized by McSweeney (2002).

CONCLUSION

It is difficult to make comparisons between self-efficacy beliefs because of the strong relationship between cultures and self-efficacy beliefs at local, national and international levels. Micro and macro contexts in which both science teachers and students form their beliefs vary substantially from one classroom to the next, and both within schools and between countries. For example, variations in the nature of formative self-efficacy in three different countries with seemingly similar science course objectives, shape outcome efficacies in a variety of ways. These variants, 'degree of dimensionality of the school', 'feedback directness' and 'feedback transparency' can all alter outcome efficacies both for science teachers and pupils, making direct and simple comparisons between classrooms difficult (Little, et al., 2001).

Other significant cultural differences between schools and national groups such as degrees of social inequality, the role of the individual in groups, concepts of gender and ways of dealing with uncertainty also affect the formation of self-efficacy beliefs (Hofstede et al., 2010). Discovering how these patterns are expressed through the interactions of teachers and students in a classroom is an essential part of learning about how variations in self-efficacy are associated with teaching and learning science.

An important implication for researchers is that the instruments used to assess self-efficacy beliefs need to be able to adequately capture these contextual variables of teaching and learning environments, rather than just asking questions about capability. Understanding the cultural norms that shape the manner and circumstances in which belief inducing teacher-student interactions occur can explicate the causes of changes in self-efficacy beliefs.

Concomitantly, self-efficacy assessment instruments should be administered so that teachers and students are assured of the anonymity of their responses. Without such assurances, they may underrate their initial competency beliefs to protect themselves against having claimed too high of an initial ability. Later, after activities intended to boost self-efficacies, when teachers and students again rate themselves, they may be more confident at claiming their ability. Consequently, the before and after comparisons of their efficacy beliefs will show inflated gains.

Useful future research can go beyond basic measures of gains in efficacy by parsing out the relative contributions of each of Bandura's (1997) four ways to alter self-efficacy. For instance, it would be worth exploring the distribution of an increase in self-efficacy for conducting science experiments among *mastery experiences* and *verbal persuasion*. Such fine-grained research would more naturally be considerate of cultural differences, since it would require close and revealing examinations of precisely which factors are efficacious in changes in self-efficacy.

REFERENCES

Bandura, A. (1982). The assessment and predictive generality of self-precepts of efficacy. *Journal of Behavioral, Theoretical and Experimental Psychiatry, 13*, 195–199.
Bandura, A. (1997). *Self-efficacy: The exercise of control.* New York, NY: Freeman.
Enochs, L. G., & Riggs, I. M. (1990). Further development of an elementary science teaching efficacy belief instrument: A pre-service elementary scale. *School Science and Mathematics, 90*(8), 694–706.
Geertz, C. (1994). Thick description: Toward an interpretive theory of culture. In M. Martin & L. McIntyre (Eds.), *Readings in the philosophy of social science.* Boston, MA: Massachusetts Institute of Technology Press.
Gibson, S., & Dembo, M. H. (1984). Teacher efficacy: A construct validation. *Journal of Educational Psychology, 76*, 569–582.
Hofstede, G. H. (1991). *Cultures and organizations: Software of the mind.* Berkshire, United Kingdom: McGraw-Hill Book Company Europe.

Hofstede, G. H., Hofstede, G. J., & Minkov, M. (2010). Cultures and organizations: Software of the mind: Intercultural cooperation and its importance for survival. New York, NY: McGraw-Hill.

Little, T. D., Lopez, D. F., Oettingen, G., & Baltes, P. B. (2001). A comparative-longitudinal study of action-control beliefs and school performance: On the role of context. *International Journal of Behavioral Development, 25*(3), 237–245.

Little, T. D., Oettinger, G., Stetsenko, A., & Baltes, P. B. (1995). Children's action-control beliefs about school performance: How do American children compare with German and Russian children? *Journal of Personality and Social Psychology, 69*(4), 686–700.

Marx, D. M., Ko, S. J., & Friedman, R. A. (2009). The Obama effect: How a salient role model reduces race-based performance differences. *Journal of Experimental Social Psychology, 45*(4), 953–956.

McSweeney, B. (2002). Hofstede's model of national cultural differences and their consequences: A triumph of faith – A failure of analysis. *Human Relations, 55*, 89–118.

Organisation for Economic Co-operation and Development (OECD). (2010). *Programme of International Student Assessment (PISA) 2009 Results: What Students Know and Can Do – Student Performance in Reading, Mathematics and Science (Volume I)*. Retrieved from http://dx.doi. org/10.1787/9789264091450-en

Oettingen, G. (1995). Cross-cultural perspectives on self-efficacy. In A. Bandura (Ed.), *Self-efficacy in Changing Societies* (pp. 149–176). Cambridge, UK: Cambridge University Press.

Provasnik, S., Kastberg, D., Ferraro, D., Lemanski, N., Roey, S., & Jenkins, F. (2012). *Highlights from Trends in International Mathematics and Science Study (TIMSS) 2011: Mathematics and Science Achievement of U.S. Fourth- and Eighth-Grade Students in an International Context* (NCES 2013-009). Washington, DC: National Center for Education Statistics, Institute of Education Sciences, U.S. Department of Education.

Riggs, I. M., & Enochs, L. G. (1990). Toward the development of an elementary teacher's science teaching efficacy belief instrument. *Science Education, 74*, 625–637.

Vieluf, S., Kaplan, D., Klieme, E. & Bayer S. (2012), *Teaching Practices and Pedagogical Innovation: Evidence from the Teaching Practices and Pedagogical Innovation Study (TALIS), OECD Publishing*. Retrieved from http://dx.doi.org/10.1787/9789264123540-en

Weinburgh, M. (2007). The effect of *Tenebrio obscurus* on preservice teachers' content knowledge, attitudes, and self-efficacy. *Journal of Science Teacher Education, 18*, 801–815.

AFFILIATIONS

Robert H. Evans
Department of Science Education
University of Copenhagen

ANDREW LUMPE, AMY VAUGHN,
ROBIN HENRIKSON & DAN BISHOP

TEACHER PROFESSIONAL DEVELOPMENT AND SELF-EFFICACY BELIEFS

INTRODUCTION

While school effectiveness recommendations address a litany of factors, it is becoming increasing clear that teachers are critically important to the success of education reforms since they play such a key role in directly impacting student learning (Borko, 2004; Nye, Konstantopoulos & Hedges, 2004; Fullan, Hill & Crevola, 2006). The Teaching Commission (2004) stated that teachers are "our nation's most valuable profession" (p.12). In light of the increasing emphasis placed on the teacher and their professional actions, some reform efforts began to focus on improving teacher quality. Teacher variables related to the improvement of student learning include many factors related to professional competence and practice. One critical variable related to student learning is the teacher's salient beliefs about their teaching effectiveness (Lumpe, Czerniak, Haney & Beltyukova, 2011).

Teacher professional development as a means toward developing quality teachers is cited as a significant variable in determining school policy, setting classroom practices, and ultimately impacting student learning (Borko, 2004, Desimone, Smith & Frisvold, 2007; Smith, Desimone & Ueno, 2005; Desimone, Smith, Hayes & Frisvold, 2005). Loucks-Horsley, Love, Stiles, Mundry & Hewson (2003) maintain that teacher belief systems must be a component of teacher professional development; therefore, the main theses of this chapter are that teacher quality impacts student learning, involves belief systems, can be improved through professional development, and teacher beliefs should be a target of professional development. Our goal is to systematically connect these ideas and provide practical recommendations for designing professional development in order to maximize teacher beliefs for improving student learning.

EFFECTIVE TEACHER PROFESSIONAL DEVELOPMENT

Teachers are adults and their professional learning cannot be treated the same as the learning of children. Lindeman (1926) first broached the subject when he acknowledged that approaches to adult education should be through contexts and situations, not subjects. When students learn, they adapt to an established curriculum;

R. Evans et al., (Eds.), The Role of Science Teachers' Beliefs in International Classrooms, 49–63.
© 2014 Sense Publishers. All rights reserved.

whereas when teachers (or adults) learn, the curriculum is built around their needs. Lindeman stated, "Subject matter is brought into the situation, is put to work, when needed . . . The situation-approach to education means that the learning process is at the outset given a setting of reality" (p. 15). Knowles, Holton & Swanson (2005) followed up Lindeman's theory of adult learning with the concept of andragogy, or adult learning. They describe andragogy for adults who want to self-direct their own learning, who are interested in growing in their knowledge, and are autonomous learners. Knowles refers to it as "the art and science of helping adults learn" (Knowles, 1973, p. 40). Adult learners require opportunities to direct their own learning, use life experiences that enhance their learning, and apply their learning immediately in problem-centered situations (Knowles, Holton & Swanson, 2005; Merriam, 2001). Adult learning is a cultural event grounded in experiences and belief systems.

Bandura (1989, 2005) addressed important aspects of adult learning that include the influence of motivation and how it affects both individuals and groups and individuals' self-efficacy and how that may influence collaboration (Gredler, 2009). These ideas are important when considering how adults learn together leading to the design of professional development to maximize the learning that takes place within a group of professional adults such as teachers.

Many traditionally employed forms of professional development, such as single session workshops, are woefully inadequate (Borko, 2004; Barber & Mourshed, 2007; Lumpe, 2007; Darling-Hammond, Chung Wei, Andree, Richardson & Orphanos, 2009). Within the past decade, researchers began to reveal that effective teacher professional development, as a form of adult learning, has the potential to serve as a key school organizational component for the improvement of teaching and student learning (Elmore, 2007; Fullan, 2007). Realizing its potential, Borko (2004) called for continued research to understand the key elements and interactions of effective teacher professional development.[1]

In a detailed analysis of 25 teacher professional development programs, Blank, de las Alas & Smith, (2008) found that among programs that demonstrated positive impacts on student outcomes, several features were present including over 50 hours or more of professional development, continuous coaching and mentoring for teachers, alignment of curriculum, and ongoing teacher collaboration. In a comprehensive review of research, Yoon, Duncan, Lee, Scarloss & Shapley (2007) also found that the duration of professional development demonstrated positive and significant impact on student learning.[1] Darling-Hammond et al. (2009) summarized research on teacher professional development by claiming that successful programs are sustained over time, utilize collaborative approaches, focus on the content being taught in the classroom, and provide multiple opportunities for teachers to apply what they learned.

Based on the earlier research of Garet, Porter, Desimone, Birman & Yoon (2001), Desimone (2009) proposed a model of teacher professional development that includes the following core features: 1. focus on content to be learned by students, 2. active learning during the professional development experiences, 3. coherence of the learning to teachers' professional needs, 4. occurrence over a long duration, and 5. collective participation by educators in their professional learning. When these features occur in a sustained manner, it is anticipated that teachers' knowledge, skills, beliefs, and attitudes will improve thereby increasing student learning. This type of professional development must occur in the context of a supportive school system that includes curriculum, leadership, and policy. Guskey (2002) proposed a similar model where effective professional development would lead to improved professional practices, which would in turn lead to improvements in student learning. While both Desimone and Guskey cite the importance of teacher beliefs and attitudes in the professional development process, Desimone directly emphasized their importance by including these affective variables as part of the development process where Guskey lists them as outcomes after student learning occurs.[1]

Loucks-Horsley et al. (2003) proposed a professional development model for science and mathematics which addresses two primary factors impacting teachers' goals, plans, and actions in the classroom. One of these factors is the context of the teaching environment – equipment, materials, physical space, support systems, etc. The other factor is the beliefs of teachers about their professional actions. They noted that self-reflection by teachers provides a feedback loop for goal modification. For example, one cannot simply give quality science curriculum materials to a teacher and expect quality science instruction. Their model supports Haney and Lumpe's (1995) contention that science teachers' beliefs must be identified and clarified prior to, and during professional development activities.[1]

In addition to a myriad of process factors related to effective professional development, all of the models described above identify the critical nature of the motivational beliefs of teachers. According to Bandura (1997), beliefs are thought to provide the best indication of the decisions people make throughout their lives. Teachers possess beliefs regarding their professional practice that may, in turn, impact student learning. The learning provides a feedback loop in modified beliefs.[1] An outline of this proposed feedback loop is shown in Figure 1. Researchers recently began to document this connection between teacher beliefs and student learning (e.g., Lumpe et al., 2011; Goddard, Hoy & Hoy, 2004; Ross, Hogaboam-Gray & Hannay, 2001).

Background on Teacher Beliefs

In his Motivation Systems Theory, Ford (1992) argued that competence in any given area, such as effective science teaching, is a combination of a person's motivation, skill, and environment. Motivation, he further clarified, is composed of goals and personal agency beliefs. Goals "are thoughts about desired states or outcomes that one would like to achieve" (p. 248). For teaching, goals could include planning lessons, using effective instructional strategies, positively impacting student learning and relating other components to teaching. According to Ford, personal agency beliefs are evaluative beliefs comparing a person's goals with the consequences of their pursuit of those goals.[1] Ford (1992) stated,

51

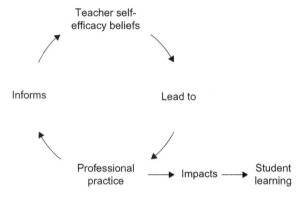

Figure 1. Beliefs/Actions Feedback Loop

. . . personal agency beliefs play a particularly crucial role in situations that are of the greatest developmental significance-those involving challenging but attainable goals. Consequently, they are often key targets of intervention for parents, teachers, counselors, and others interested in promoting effective functioning. (p. 124-125)

Ford's theory identifies two types of personal agency beliefs, capability and context. Capability beliefs are synonymous with Bandura's (1997) concept of self-efficacy. Capability beliefs are beliefs about one's ability or skill to meet a particular goal. This could be characterized as a teacher's belief that they can effectively teach science to children. Bandura further delineated a related belief construct called outcome expectancy. He stated, " . . . outcome expectation is a judgment of the likely consequences such performances will produce" (p. 21). For a schoolteacher, this would be a belief that if one teaches effectively, then students will learn.[1]

Context beliefs are beliefs about the responsiveness of the environment including external factors and people. Context beliefs are sometimes called perceptions of control. They are similar to Ajzen and Madden's (1986) perceived behavioral control construct and Bandura's (1997) outcome expectancy construct. Context beliefs include the role of the entire context in meeting desired goals. In the case of teaching, context beliefs would not only encompass the students but also administrators, parents, other teachers, institutions, organizations, and the physical environment. In education, contexts can be broadly classified into the designed environment (e.g., buildings, equipment), human environment (e.g., students, faculty, parents), and socio-cultural environment (e.g., policy, cultural norms) (Ford, 1992).[1] Lumpe, Haney & Czerniak (2000) found that a group of science teachers possessed fairly positive context beliefs which should allow them to function effectively in the classroom.

In his social learning cognitive theory, Bandura (1989) explains how people determine their own activities and actions through a combination of both their own motivations as well as pre-determined environmental influences. At the core of Bandura's theory (2005) is the notion that individuals form action plans based on forethought which guide future activities. This action plan is based not only on a combination of both an individual's motivation and desires, but also the structures predetermined by one's environment. People also utilize self-awareness and reflection based on prior experiences in order to help them make future decisions.

The system in which people live does not completely dictate a person's actions nor does an individual have complete autonomy over their actions. Bandura (2005) makes the point that "self-generated influences" (p. 1175), are a contributing factor in human action. Bandura uses this argument when discussing how an individual's self-efficacy influences both their actions as well as their environment, including how collective group efficacy can also influence the actions of a group. Teachers usually work in groups related to a school building, grade level, or discipline.

Many challenges teachers face today require group efforts to overcome. Bandura (1982) proposed that self-efficacy affects individuals and their willingness to work with others toward improvement. Those with a stronger sense of self-efficacy are often times more willing to put forth effort toward problem solving and persevere longer than those who do not. He then connects one's individual efficacy to the collective efficacy of a group. It is the strength of a group that enables people to solve problems and enact change (Bandura, 1982). These changes, as Bandura continues to explain, call for a greater commitment from individuals sharing a common purpose; it requires effective action that merges the self-interests of those involved to support-shared goals. When groups have individuals with a weak sense of self-efficacy, it in turn affects the efficacy of the group. Challenges arise that interfere with the general force of the group. Bandura's ideas reflect the challenges that teachers have today when attempting to work together toward a common goal of collective professional learning. Individuals play an important role in the success or failure of the community of learning based on the attitudes, beliefs, and knowledge they bring with them.

In practice, Bandura (1989) argued that people predict outcomes of their actions based on their previous knowledge and experiences. While those who believe strongly in their problem-solving skills are more efficient in their abilities to think through difficult situations, those who have a lower self-efficacy tend to be unable to think through potential outcomes in any organized way. This same theory holds true to the collective group efficacy. When groups have a strong sense of their abilities to think through possible scenarios, they are able to organize and prepare for future encounters. Bandura also notes that even though there may be individuals who have a high degree of self-efficacy, the efficacy of the group is low due to the inability to work together as a team (2000, p. 75); therefore group efficacy involves improving the collective body rather than the individuals.

Within a school environment, teachers engage in both group and individual problem solving and make decisions based on prior knowledge. As Bandura (1993) writes:

[s]chools in which the staff collectively judge themselves as powerless to get students to achieve academic success convey a group sense of academic futility that can pervade the entire life of the school. School staff members who collectively judge themselves capable of promoting academic success imbue their schools with a positive atmosphere for development. (p. 141)

Following these ideas, the weaker the perceived self and group efficacy, the less organized and thorough the actions of that individual or group will be to enact any change for students or the school system. In effect, people create their surrounding environments (Bandura, 2000). Within schools, teachers have the ability to collectively influence and create the environments in which they work thus creating not only a collective group efficacy but also a norm for operating at the school level.

Bruner (1996) also discussed learning based on life experiences as four crucial ideas that include agency, reflection, collaboration, and culture. Agency is the ability to take control of one's thinking and to create learning through experiences. It is similar to Ford's (1992) notion of personal agency beliefs and Bandura's (1997) definition of self-efficacy. Reflection is making sense of new learning and understanding it. Through collaboration, this new learning can be shared and utilized by others. Finally, culture is the environment that the participants construct and work within. Collaborative groups may have great potential to stimulate these four crucial components. Teachers use their own experiences when meeting collaboratively. They learn and reflect on these classroom experiences in a collaborative way with other professionals and improve their practice through this dialogue, reflection, and learning.

Since teacher beliefs serve as one important factor in the goal of teaching effectiveness, they recently became the target of professional development programming. In a study of a teacher professional development program, Rosenfeld and Rosenfeld (2008) found that mediated professional development activities caused an increase in teacher beliefs about student learning. They concluded that teacher professional development should be linked to teacher belief systems. Ross and Bruce (2007) conducted a comparison of teacher professional development programs on teacher's efficacy. They found that treatment teachers outperformed control group teachers on one measure of efficacy. They suggested that researchers examine connections between teacher beliefs and student achievement.[1]

Because of increasing calls for school accountability, an increased emphasis placed on the role of the teacher, the role of professional development, and theoretical connections between teacher beliefs and classroom action, teacher professional development programs have the potential to impact teacher belief systems, teaching practices, and student learning.[1]

ANTECEDENTS OF TEACHER BELIEFS AS RELATED TO PROFESSIONAL PRACTICE

Bandura (1997) maintained that the performance of complex tasks, such as classroom teaching, is controlled by cognitive and self-regulative processes that involve self-efficacy beliefs. He asserted that to increase one's self-efficacy, particularly a low self-efficacy, "requires explicit, compelling feedback that forcefully disputes the preexisting disbelief in one's capabilities" (p. 82). In other words, the self-efficacy beliefs of teachers can be targeted through professional development experiences. Bandura suggests the following four methods for increasing self-efficacy: enactive mastery experiences, vicarious experiences, verbal persuasion, and affective states. By addressing Bandura's methods for increasing self-efficacy in teacher professional development programs, teacher self-efficacy, and therefore student achievement, may be improved.

Mastery Experiences

People come to performance tasks with preexisting beliefs about their capabilities to perform (Bandura, 1997). When a teacher steps into the classroom, he/she has some level of belief, positive or negative, about his/her effectiveness to help children learn. These preexisting beliefs will influence the person's ability to perform the task. If teachers judge themselves as ineffective, they are more likely to engage in ineffective professional behaviors in the classroom. Conversely, positive beliefs tend to lead to more effective performance. Typically, these belief systems are resistant to change. Teachers hold fast to their long-practiced instructional strategies. Only as they acquire and successfully implement new skills do their beliefs systems begin to change.

Bandura argued that when a person experiences success or mastery, they are more likely to develop positive belief systems. As applied to teaching, mastery experiences are activities in which the teacher personally experiences a successful instructional performance ultimately leading to gains in student learning. According to Bandura, mastery experiences are the most powerful sources of efficacy information as they provide the most authentic evidence that a person can experience success when faced with challenging tasks.

Change to a teacher's self-efficacy, however, requires more than just completion of a mastery experience. The teacher must self-monitor or reflect upon the mastery experience to evaluate its implications. The extent to which the teacher will alter his or her perceived self-efficacy is dependent on many factors. These include, but are not limited to, the teacher's appraisal of his or her ability, the perceived difficulty of the task, the amount of effort expended on the task, and the previous pattern of success and failure (Bandura, 1997). Critical reflection, as a form of metacognition, occurs when learners construct their own narratives based on learning experiences and professional practice (Ellis, 2001). Schön's (1983) seminal work on professional reflective practice paved the way to address this topic in teacher education. As applied to teaching, approaches that support the examination of beliefs that emerge from these practices promote the development of more flexible and intentional approaches to effective teaching and learning (Sockman & Sharma, 2008; Schoffner, 2009).

A mastery experience may be achieved through scaffolding, as the environment may be intentionally structured so that the teacher can perform successfully despite his or her weaknesses. This treatment provides opportunities for the teacher to make corrective refinements toward the perfection of skills (Bandura, 1977). Such intentional structuring may include practical opportunities for a teacher to try newly learned strategies in their classroom.

Given the complex nature of the teaching enterprise and teachers' preexisting and oftentimes resistant beliefs, it may take a long time to make a positive impact on beliefs and skills. This confirms Desimone's (2009) assessment that teacher professional development must occur over a period of time and avoid quick workshop configurations. Mastery is a slow, intentional, and reflective process.

Vicarious Experiences

Vicarious experiences are activities in which a person witnesses others executing challenging tasks without adverse consequences. Modeling is the primary means through which a teacher may engage in a vicarious experience and is an effective tool for promoting self-efficacy. Vicarious experiences have more influence when the observer can identify and connect with the struggles the master teacher has overcome. For example, when a 3rd grade teacher can witness another 3rd grade teacher successfully teach a science lesson, it provides believability and evidence that the observers can also improve if they expend more energy and persist in their efforts (Bandura, 1977).

In order to be most effective, modeling should occur in diverse settings where different people demonstrate success using diverse strategies. As a teacher observes multiple examples of successful teaching, they stand a better chance of improving their own confidence.

It is not only important for the teacher to witness a mastery experience performed by a person of similar teaching background and experience, the teacher must also be aware of the processes that the performer engaged in internally. According to Bandura (1997),

> [i]t is difficult to acquire cognitive skills through modeling when covert thought processes are not adequately reflected in modeled actions. The problem of observability is overcome simply by having models verbalize their thought processes and strategies aloud as they engage in problem-solving activities. The covert thought guiding the actions are thus made observable through overt representation. (p. 93)

Reflective discussions during or after the modeling process can reveal the reasons for various actions and consequences. For example, master teachers may discuss their actions with a group of teachers while watching a video of their teaching actions. Only then can teachers attempt and successfully achieve mastery for themselves.

Verbal Persuasion

Belief systems can be positively influenced if others can tell their success stories and use logical arguments for effective practice. Verbal persuasion is widely used because of its ease and availability. Through verbal persuasion, teachers are led, through suggestion, into believing that they can perform successfully in situations that have overwhelmed them in the past (Bandura, 1977). This may occur when effective teachers explain the strategies used to successfully teach children and can demonstrate the impact on student learning. These self-affirming beliefs promote effort, the development of skills, and a sense of personal efficacy.

Verbal persuasion can also be conveyed to teachers through evaluative feedback. The evaluative feedback, however, must be constructive and framed so as not to undermine the teacher's sense of efficacy (Bandura, 1997). Such systems provide an opportunity for formative processes that include feedback and reflection on professional practices. This can serve as a form of metacognition for teachers (Flavell, 1979). The move to teacher evaluation systems based on effective instructional strategies and coupled with constructive feedback on actual performance, can serve to build confidence in teaching capability (Darling-Hammond, 2010).

Affective States

Affective states also impact teachers' appraisal of their self-efficacy as mood may bias how events are interpreted. A positive mood enhances perceived efficacy; whereas despondency and stress diminishes it. If teachers feel as though they are not effective, it may lead to a self-fulfilling prophecy. Confidence tends to breed success. It is not the affective state itself that is critical to effective functioning; it is the cognitive processing of the physiological state and mood that makes the difference in how one perceives effectiveness. As Bandura states,

> "People then act in accordance with their mood-altered efficacy beliefs, choosing more challenging tasks in a self-efficacious frame of mind than they do when they doubt their efficacy . . . [B]y raising efficacy beliefs that heighten motivation and performance accomplishments, good mood can set in motion an affirmative reciprocal process." (Bandura, 1997, p. 113)

Teachers need to be aware of their stress levels and mood during teaching. They should be able to reflect upon the mitigating factors that cause stress and be able to make adjustments when needed.

As related to teacher development, mastery experiences may include the identification of preexisting beliefs, experiencing success in the classroom, reflection on professional practices, all occurring over long duration. Vicarious experiences rely on modeling in multiple settings coupled with reflection. Verbal persuasion involves logical arguments for a program, telling success stories, and positive performance evaluation. Affective experiences include a reduction of stress and building of confidence. The four antecedents of self-efficacy beliefs are not mutually exclusive and they may occur simultaneously. These experiences should be purposefully built into teacher professional development activities. An example of a teacher professional development program is included in the next section in order to demonstrate how self-efficacy beliefs can be integrated into teachers' experiences.

PROFESSIONAL DEVELOPMENT STRATEGIES TO IMPACT TEACHER BELIEFS

A teacher project funded by the National Science Foundation (NSF) is used to demonstrate how teacher belief systems can be targeted during professional development. The project was called the Toledo Area Partnership in Education: Support Teachers as Resources to Improve Elementary Science (TAPESTRIES) and was funded under the NSF's Local Systemic Initiative program in the late 1990s. The long-term project included goals that aimed to improve science teaching and learning at the elementary level. The project was a direct application of the science teacher professional development models from Loucks-Horsley et al. (2003) and Haney and Lumpe (1995). A previously published report about this project (Czerniak, Beltyukova, Struble, Haney & Lumpe, 2006) provides great detail about the project activities. A brief description of components of the program that played a critical role in the implementation of systematic science reform is provided below. Wherever program components take into account Bandura's (1997) four antecedents of effective belief systems for effective functioning (mastery experience, vicarious experience, verbal persuasion, and affective states), it is noted within the project description in parentheses.

Six, two-week-long Summer Institutes for classroom teachers were conducted each year of the project. Teachers participated in sessions that focused on inquiry-based instruction, science content knowledge, and science process culled from the districts' adopted curriculum (mastery experience). The adopted curriculum included the Full Option Science System (FOSS) developed by the University of California's Lawrence Hall of Science and Science, Technology, and Children (STC) developed by the National Science Resources Center (NSRC). The Summer Institutes ran eight hours a day for two weeks for approximately total of 80 contact hours (mastery experience). The summer institutes were co-taught by science educators, full time Support Teachers, and university scientists. The sessions followed the inquiry-based 5E learning cycle model (Bybee & Landes, 1988) while simultaneously introducing pedagogy and science content (vicarious experience). During the sessions, the leaders facilitated sessions on the effectiveness of the instructional materials and teaching strategies on children's learning (verbal persuasion). One goal was to build confidence and skill in using the instructional materials as the teachers were preparing to implement them during the upcoming academic year (affective states).

Support Teachers, elementary teachers who were given full time release from teaching responsibilities, provided assistance to classroom teachers implementing science inquiry, helped teachers with district assessments, and executed their district action plans for improving science literacy (mastery experiences, vicarious experiences, verbal persuasion). Support Teachers received more than 200 contact hours of leadership training in the form of a two-week Summer Institute, two graduate courses, a staff retreat, and a spring conference (mastery experiences).

Professional development was sustained during the academic year by focusing on the implementation of the adopted curriculum and assessments (mastery experiences). Depending on the grade level, science was typically taught 3-4 days per week for approximately 45-60 minutes each day. The Support Teachers visited an assigned cohort of teachers biweekly and provided assistance with science curriculum preparation, gave strategies for teaching science, supplied science content background information with the help of the university scientists, assisted with classroom and district science performance-based assessments, modeled science lessons, and offered peer coaching for the classroom teacher (mastery experiences, vicarious experiences, verbal persuasion). When it became apparent that the mandated reading and writing assessments were considered more important, an emphasis was made to integrate science into basic literacy in order to reduce stress levels of the teachers (affective states). Each teacher conducted a "research lesson" – a Japanese-style lesson study that involved the teacher writing a lesson in the 5E learning cycle model and writing a two-page reflective analysis of the lesson identifying specific strengths and weaknesses (mastery experiences). These academic year activities provided approximately 24 additional hours of professional development spread evenly over the academic year.

All principals participated in a one-day retreat and follow-up sessions throughout the academic year. Model lessons were presented, and principals were made aware of science education reform research (vicarious experiences, verbal persuasion). Additionally, the project leaders solicited principal support for the project and their input on the challenges of implementing science reform (affective states). Support Teachers scheduled two local community meetings to involve city leaders, parents, and local principals in this science reform effort. These meetings took many forms - i.e., family science days, parent-teacher organization meetings, and state achievement test information sessions (verbal persuasion).

In a set of research studies about this project, Czerniak et al. (2006) demonstrated the impact of the professional development activities on student achievement when they tracked state science achievement test scores for over 8,000 students across 43 elementary school buildings. Statistical comparisons were made on test scores from before and during implementation of the project. It was found that when controlling for prior achievement, science test scores improved significantly after implementation of the project. A cumulative effect was also found - students who had consecutive years of trained teachers outperformed students who were not exposed to multiple years of trained teachers. School buildings with greater degrees of project implementation in terms of contact hours outscored schools with less involvement.

In an additional study associated with this project, Lumpe et al. (2011) explored relationships between teachers' beliefs and student achievement. They found that elementary teachers who participated in long-term, intense professional development program in science showed significant gains in their science teaching self-efficacy. Ross and Bruce (2007) also found that professional development had a significant effect on one dimension of teacher self-efficacy. Furthermore, teacher self-efficacy beliefs were found to be a significant and positive predictor of student achievement (Lumpe et al., 2011; Ross et al., 2001).

CONCLUSION

Well-designed and implemented professional development activities have the potential to positively impact teachers' beliefs about teaching science. Theoretically (Ford, 1992; Lumpe, Haney & Czerniak, 2000; Goddard, Hoy & Hoy, 2004), such beliefs may play a role in the quality of teaching and ultimately, student learning. The approaches used in the program described above were designed to identify and positively impact teachers' beliefs. These approaches included opportunities to try new teaching strategies in their classrooms while receiving high levels of support.[1] The professional development activities purposefully addressed Bandura's (1997) four principal sources of beliefs - mastery experiences, vicarious experiences, verbal persuasion, and affective states. Each of these sources may purposefully be targeted in order to develop positive and robust beliefs leading to more effective teacher functioning in the classroom.[1] Mastery experiences for teachers may include opportunities to apply effective teaching strategies, reflect upon and modify teaching practices, and ultimately experience student-learning success. Observing successful master teachers as a form of modeling can provide vicarious learning experiences, and engaging in logical discussions about standards, curriculum, and effective instruction may help persuade teachers to align their teaching with policy and research-based recommendations. Teacher's affective states about teaching science, or emotional tone, may be impacted by a variety of factors including resources, support systems, and reward systems.

The sources of effective functioning defined by Bandura are inherent in current teacher professional learning community models (DuFour, 2005). Many professional learning community protocols exist including peer coaching/teacher leadership (Reeves, 2006; Danielson, 2006), Collaborative Analysis of Student Work or CASL (Langer, Colton, & Goff, 2003) Critical Friends Groups (http://www.nsrfharmony. org/faq.html), and Japanese Lesson Study (Lewis, Perry, & Murata, 2006). These protocols utilize similar strategies including reflective inquiry, social norm setting among professionals, using student assessments to target learning gaps, and modifying instruction to address the identified gaps.[1] Those who develop professional development programs and providers should carefully prepare and present programs for teachers. Principles of effective teacher professional development proposed by Desimone (2009) and Loucks-Hoursley et al. (2003) could be used as frameworks when designing programs that may impact teacher knowledge, skills, beliefs, and ultimately student learning. Funding agencies should prioritize programs that utilize proven professional development methods that target teacher belief systems. Only then can the great expenditures be justified and ultimately student achievement impacted positively.

ACKNOWLEDGEMENTS

This research is supported in part by funding from the National Science Foundation (NSF), project no. 9731306. The views expressed here are not necessarily those of NSF.

NOTES

[1] Portions of some paragraphs were adapted from the previously published article listed below. Used with permission per retained author rights from Taylor and Francis Publishers.
Lumpe, A., Czerniak, C., Haney, J., & Beltyukova, S. (2012). Beliefs about teaching science: The relationship between elementary teachers' participation in professional development and student achievement. International Journal of Science Education, 34(2), 153-166.

REFERENCES

Ajzen, I., & Madden, T. J. (1986). Prediction of goal-directed behavior: Attitudes, intentions, and perceived behavioral control. *Journal of Experimental Social Psychology, 22,* 453–474.

Bandura, A. (1977). Self-efficacy: Toward a unifying theory of behavioral change. *Psychological Review, 84*(2), 191–215.

Bandura, A. (1982). Self-efficacy mechanism in human agency. *American Psychologist, 37*(2), 122–147.

Bandura, A. (1989). Human agency in social cognitive theory. *American Psychologist, 44*(9), 1175–1184.

Bandura, A. (1993). Perceived self-efficacy in cognitive development and functioning. *Educational Psychologist, 28*(2), 117–148.

Bandura, A. (1997). *Self-efficacy: The exercise of control.* New York, NY: W. H. Freeman and Company.

Bandura, A. (2000). Exercise of human agency through collective efficacy. *American Psychological Society, 9*(3), 75–78.

Bandura, A. (2005). The evolution of social cognitive theory. In K. G. Smith & M. A. Hitt (Eds.), *Great Minds in Management* (pp. 9–35) Oxford: Oxford University Press.

Barber, M., & Mourshed, M. (2007). *How the world's best-performing school systems come out on top.* McKinsey & Company.

Blank, R. K, de las Alas, N., & Smith, C. (2008). *Does teacher professional development have effects on teaching and learning? Analysis of evaluation finding from programs for mathematics and science teachers in 14 states.* Washington, DC: Council of Chief State School Officers.

Borko, H. (2004). Professional development and teacher learning: Mapping the terrain. *Educational Researcher, 33*(8), 3–15.

Bruner, J. (1996). *The culture of education.* Boston, MA: Harvard University Press.

Bybee, R. W., & Landes, N. M. (1988). What research says about the new science curriculum (BSCS). *Science and Children, 25,* 35–39.

Czerniak, C. M., Beltyukova, S., Struble, J., Haney, J. J., & Lumpe, A. T. (2006). Do you see what I see? The relationship between a professional development model and student achievement. In R. E. Yager (Ed.), *Exemplary science in grades 5-8: Standards-based success stories* (pp. 13–43). Arlington, VA: NSTA Press.

Darling-Hammond, L., Wei, C. R., Andree, A., Richardson, N., & Orphanos, S. (2009). *Professional learning in the learning profession: A status report on teacher development in the United States and abroad.* Dallas, TX: National Staff Development Council and School Redesign Network at Stanford University.

Darling-Hammond, L. (2010). *Evaluating teacher effectiveness: How teacher performance assessments can measure and improve teaching.* Washington, DC: Center for American Progress.

Desimone, L. M., Smith, T. M., Hayes, S., & Frisvold, D. (2005). Beyond accountability and average math scores: Relating multiple state education policy attributes to changes in student achievement in procedural knowledge, conceptual understanding and problem solving in mathematics. *Educational Measurement: Issues and Practice, 24*(4), 5–18.

Desimone, L. M., Smith, T., & Frisvold, D. (2007). Is NCLB increasing teacher quality for students in poverty? In A. Gamoran (Ed.), *Standards-based and the poverty gap: Lessons from no child left behind* (pp. 89–119). Washington, DC: Brookings Institution Press.

Desimone, L. M. (2009). Improving impact studies of teachers' professional development: Toward better conceptualizations and measures. *Educational researcher, 38*(3), 181–199.

DuFour, R. (2005). What is a professional learning community? In R. DuFour, R. Eaker, & R. DuFour (Eds), *On common ground* (pp. 31–43). Bloomington, IN: Solution Tree.

Ellis, A. K. (2001). *Teaching, learning, and assessment together: The reflective classroom.* Poughkeepsie, NY: Eye on Education.

Elmore, R. F. (2007). *School reform from the inside out: Policy, practice, and performance.* Cambridge, MA: Harvard Education Press.

Flavell, J. H. (1979). Metacognition and cognitive monitoring: A new area of cognitive-developmental inquiry. *American Psychologist, 34*(10), 906–911.

Ford, M. E. (1992). *Motivating humans: Goals, emotions, and personal agency beliefs.* Newbury Park, CA: Sage.

Fullan, M., Hill, P., & Crevola, C. (2006). *Breakthrough.* Thousand Oaks, CA: Corwin Press.

Fullan, M. (2007). *The new meaning of educational change* (4th ed.). New York, NY: Teachers College Press.

Garet, M. S., Porter, A. C., Desimone, L., Birman, B. F., & Yoon, K. S. (2001). What makes professional development effective? Results from a national sample of teachers. *American Educational Research Journal, 38*(4), 915–945.

Goddard, R. D., Hoy, W. K., & Hoy, A. W. (2004). Collective efficacy beliefs: Theoretical developments, empirical evidence, and future directions. *Educational Researcher, 33*(3), 3–13.

Gredler, M. (2009). *Learning and instruction: Theory into practice.* Upper Saddle River, NJ: Pearson.

Guskey, T. T. (2002). Professional development and teacher change. *Teachers and teaching: Theory and practice, 8*(3/4), 381–391.

Haney, J. J., & Lumpe, A. T. (1995). A teacher professional development framework guided by science education reform policies, teachers' needs, and research. *Journal of Science Teacher Education, 6*(4), 187–196.

Knowles, M. (1973). *The modern practice of adult education. From pedagogy to andragogy.* Englewood Cliffs, NJ: Cambridge.

Knowles, M., Holton, E., & Swanson, R. (2005). *The adult learner: The definitive classic in adult education and human resource development.* San Diego, CA: Elsevier.

Langer, G. M., Colton, A. B., & Goff, L. S. (2003). *Collaborative analysis of student work: Improving teaching and learning.* Alexandria, VA: Association for Supervision and Curriculum Development.

Lewis, C. C., Perry, R., & Murata, A. (2006). How should research contribute to instructional improvement? The case of lesson study. *Educational Researcher, 35*(3), 3–14.

Lindeman, E. (1926). *The meaning of adult education.* New York, NY: New Republic.

Loucks- Hoursley, S., Love, N., Stiles, K. E., Mundry, S. E., & Hewson, P.W. (2003). *Designing professional development for teachers of science and mathematics.* Thousand Oaks, CA: Corwin Press.

Lumpe, A. T., Haney, J. J., & Czerniak, C. M. (2000). Assessing teachers' beliefs about the science teaching context. *Journal of Research in Science Teaching, 37*(3), 275–292.

Lumpe, A. T. (2007). Research based professional development: Teachers engaged in professional learning communities. *Journal of Science Teacher Education, 18*(1), 125–128.

Lumpe, A., Czerniak, C., Haney, J., & Beltyukova, S. (2012). Beliefs about teaching science: The relationship between elementary teachers' participation in professional development and student achievement. *International Journal of Science Education, 34*(2), 153–166.

Merriam, S. (2001). The new update on adult learning theory. Andragogy and self-directed learning: Pillars of adult learning theory. *New Directions for Adult and Continuing Education, 89*, 3–14.

Nye, B., Konstantopoulos, S., & Hedges, L. (2004). How large are teacher effects? *Educational Evaluation and Policy Analysis, 26*(3), 237–257.

Reeves, D. B. (2006). *The learning leader: How to focus on school improvement for better results.* Alexandria, VA: Association for Supervision and Curriculum Development.

Rosenfeld, M., & Rosenfeld, S. (2008). Developing effective teacher beliefs about learners: The role of sensitizing teachers to individual learning differences. *Educational Psychology, 28*(3), 245–272.

Ross, J., Hogaboam-Gray, A., & Hannay, L (2001). Effects of teacher efficacy on computer skills and computer cognitions of K-3 students. *Elementary School Journal, 10*(2), 141–156.

Ross, J., & Bruce, C. (2007). Professional development effects on teacher efficacy: Results of randomized field trial. *Journal of Educational Research, 101*(1), 50–60.

Schoffner, M. (2009). The place of the personal: Exploring the affective domain through reflection in teacher preparation. *Teaching and Teacher Education, 25*, 783–789.

Schön, D. A. (1983). *The reflective practioner.* New York, NY: Basic Books.

Sockman, B. R., & Sharma, P. (2008). Struggling toward a transformative model of instruction: It's not so easy! *Teaching and Teacher Education, 24*(4), 1070–1082.

Smith, T. M., Desimone, L. M., & Ueno, K. (2005). Highly qualified to do what? The relationship between NCLB teacher quality mandates and the use of reform-oriented instruction in middle school math. *Educational Evaluation and Policy Analysis, 27*(1), 75–109.

Yoon, K. S., Duncan, T., Lee, S. W-Y, Scarloss, B., & Shapley, K. (2007). *Reviewing the evidence on how teacher professional development affects student achievement* (Issues & Answers, REL 2007-No. 033). Washington, DC: U.S. Department of Education, Institute of Education Sciences, National Center for Education Evaluation and Regional Assistance, Regional Educational Laboratory Southwest.

AFFILIATIONS

Andrew Lumpe
School of Education
Seattle Pacific University

Amy Vaughn
Washington State Office of Superintendent of Public Instruction

Robin Henrikson
School of Education
Seattle Pacific University

Dan Bishop
School of Education
Seattle Pacific University

KADIR DEMIR & CHAD D. ELLETT

CROSS-CULTURAL RESEARCH AND PERSPECTIVES ON EPISTEMOLOGY, LEARNING ENVIRONMENTS, AND CULTURE

The purpose of this chapter is to present a working model that depicts the interactive relationships between the development of Epistemological Beliefs (EBs) of teachers and students, cultures, and learning environments (LEs). The working model recognizes how EBs can be influenced by school level and external LEs and cultures as well. Of particular interest is literature that documents cross-cultural differences in teachers' and students' EBs with a particular focus on EBs of teachers. The literature in EBs in science classrooms is a primary focus and cross-cultural studies of other teaching and learning contexts are also included. An international example, Turkey, is used to highlight the historical influence of culture on the development of various EBs.

SOME NOTES ON EPISTEMOLOGY AND EPISTEMOLOGICAL BELIEFS

Epistemology can be defined as a branch of philosophy that studies the nature of knowledge and its development (Hofer & Pintrich, 2002). A teacher's personal epistemology exerts a powerful influence on the ability to perceive and engage the diversity and complexity of LEs (Khine, 2008). How learners view knowledge and knowing has been studied under the broader heading of personal epistemology (Hofer & Pintrich, 2002). Regarding the *nature of knowledge* researchers try to understand individuals' perceptions of scientific knowledge as certain, unchanging truth, or uncertain, ever changing multiple truths.

Personal epistemology research goes back to Perry's (1970) work on the developmental stages of epistemology of college undergraduates in the USA. Since then researchers have examined the relationship between EBs and characteristics of LEs including classroom practices. This research focus is rather broad and includes cognitive outcomes, conceptual change, reflective thinking, cognitive processing, the nature of knowledge and knowing, the role and influence of an individual's beliefs, nature of knowledge on comprehension, domain specific epistemology, and teacher and student epistemologies, and meta-cognitive variables (Bendixen & Feucht, 2010). Whatever the research focus, both teacher and student EBs, and other's EBs as well (e.g., family members) are considered an important part of student learning, particularly in subjects such as science and mathematics. While there are studies comparing students' and teachers' EBs (Tsai, 2003), the literature is relatively quiet concerning the influence of students' EBs on the EBs of teachers.

R. Evans et al., (Eds.), The Role of Science Teachers' Beliefs in International Classrooms, 65–79.
© 2014 Sense Publishers. All rights reserved.

SOME NOTES ON THE ROLE OF EPISTEMOLOGY IN LES AND CLASSROOM CULTURE

Learning strategies appear to be context-dependent. Thus, learning with understanding is dependent not only on the particular ways in which content and skills are presented and experienced by learners (students and teachers), but also on the composition of the culture of the LE (Fisher & Waldrip, 1999). Some relatively recent studies have been completed that link school culture and LE characteristics. Dhindsa (2005), for example, found differences in characteristics of LEs among students in international, public, and private schools in Brunei. Additional research has been completed on the characteristics of LEs such as the amount of structure, the quality and quantity of learning tasks, and learner characteristics such as content-specific pre-conceived ideas, motivation and learning styles (Elen, Lowyck, & Proost, 1996). As well, a number of studies of LEs in a variety of cross-cultural settings, including technology enhanced LEs (Chou & Liu, 2005) have been completed. LE characteristics have also been linked to studies of epistemology in the classroom (Elen & Clarebout, 2001).

Learning entails interaction with the environment, including interactions among students and between students and the teacher. These interactions have received attention from a number of scholars in studying the relationship between EBs and LEs (Tolhurst, 2007). Several empirical studies have generated convincing evidence that EBs influence students' approaches to problem-solving, motivation, perseverance in knowledge construction, and approaches to learning (Elen & Clarebout, 2001). Given that the teacher is an important element of the total LE, and that teacher beliefs influence teaching practices, it seems reasonable in future research to assess the extent to which teacher EBs are linked to and/or influenced by the EBs of students.

There is also evidence that implies the structure of LEs can influence students' EBs. Schommer (1994) proposed that "Epistemological beliefs affect the degree to which individuals: (a) actively engage in learning, (b) persist in difficult tasks, (c) comprehend written material, and (d) cope with ill-structured domains. In each of these areas, the evidence suggests that epistemological beliefs may either help or hinder learning" (p. 302). Brownlee et al. (2001), completed a study with teacher education students which showed that students engaged in a year-long teaching program experienced more growth in sophisticated epistemological beliefs than a control group of students who were not encouraged to explicitly reflect on their EBs.

Recently, the importance of cultural factors, including teacher-student interpersonal relationships in LEs, has drawn the attention of researchers from around the world, e.g., in Australia and Asia (Fisher & Waldrip, 1999), the United States (Levy, den Brok, Wubbels, & Brekelmans, 2003) and Europe (Konings, Brand-Gruwel, & van Merrienboer, 2005), and Turkey (Kizilgunes, Tekkaya, & Sungur, 2009). Yet, as Hodson (1999) stated, changing learners' views is not easily accomplished. Students' and teachers' conceptions of learning and approaches to learning appear to be subject to individual experiences within the context of cultural norms both within and outside of classrooms. Similarly, these conceptions of teaching and learning are mediated by classroom culture as well as cultures outside of the classroom. As Konings et al. (2005) stated, teachers who conceptualize teaching as a process in which teacher knowledge is conveyed from the teacher to the learners, ignore students' intrinsic experiences and interests and inhibit deeper learning. Thus, the challenge for teachers is to promote deeper, learned-centered approaches to teaching and learning while accommodating various cultural norms, values and beliefs. The section that follows describes our working model that integrates EBs, school and classroom cultures, and LEs.

A WORKING MODEL GUIDING THE COMPLEXITY OF RESEARCH ON EBS

Our review of existing literature on the development and strengthening of EBs, LEs and classroom and school cultures led us to the development of a working model (Figure 1) that might be useful in guiding future research to better understand EBs of teachers and students. The extant research literature on EBs has focused to a large extent on the EBs of students rather than teachers, and studies of both students and teachers and their interactions are rare. With some exceptions (Tsai, 2003), there are few studies investigating the relationship between teacher and student EBs. As well, comprehensive models that include concern for multiple cultures and LEs have not been made explicit in the EB literature.

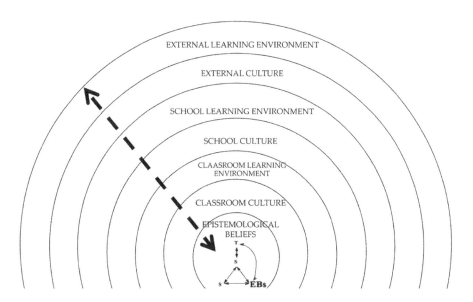

Figure 1: A Working Model Linking the Development and Strengthening of Epistemological Beliefs, Culture, and Learning Environment Characteristics (EBs- Epistemological Beliefs; S- Student; T- Teacher)

Figure 1 presents a working model that we believe can be used to better understand and articulate relationships among teacher and student EBs and multiple LEs and cultures. The model is not theory or research-based. Rather it represents an organizational framework that we think is useful given current research findings on EBs and various cultures and LEs. Figure 1 posits that multiple levels of culture and LEs and EBs are nested within and influenced by various LEs and cultures.

The model assumes *reciprocal* relationships, influences, and interactions between teacher EBs and student EBs. The model assumes that teacher and student EBs are inextricably interrelated. The proposed model also makes the following assumptions:

- Teaching and learning occur within the context of multiple LEs and multiple cultures and these collectively influence students' and teachers' EBs.
- Elements of the model are highly interactive and reciprocal in influence and each circle can influence any other circle.
- The model depicts three levels of LEs (classroom, school, and external) and three levels of culture (classroom, school, and external) that can influence teacher and student EBs.
- The heart (core) of the model consists of teacher to student and student-to-student interactions that facilitate or hinder the development and strengthening of both teacher and student EBs.
- EBs at any point in time are situated within the larger LE (classroom, school, external) and reflect multiple cultural influences as well.

The EB arrows suggest that the particular EBs students and teachers develop in a classroom are influenced by student to student and student to teacher interactions. The model assumes a constant, reciprocal ebb and flow between EBs that teachers and students bring to the classroom LE and changes in these beliefs as well. Teacher and student EBs evolve and are strengthened or weakened over time. Changes in EBs (e.g., better understanding the nature of science and the tentative nature of scientific knowledge) would be depicted by increasing the size of the inner EB circle. If EBs are weak at any point in time, the EB circle becomes smaller. Epistemological beliefs are strengthened or weakened by various elements of LEs and cultures in which these beliefs are embedded. The dotted arrow running through levels of the model indicates reciprocal interactions between EBs, multiple LEs and multiple cultures. This working model is generic in the sense that general elements of the model apply to varying contexts (e.g., countries, subject matter, grade levels, urban vs. rural schools).

The model suggests several possibilities for exploring future research such as:

- To what extent do classroom culture and classroom LE contribute to the development and strengthening of teachers' and students' EBs?
- To what extent are students' EBs developed and strengthened (or weakened) by prior knowledge and the cultural understandings, values, and beliefs they bring to the classroom?

- What actions (if any) should be taken when there are strong clashes related to EBs between the classroom culture and external cultures (e.g., family beliefs about intelligent design and evolution might clash with natural selection)?
- What is the contribution of school culture to the development of teacher EBs and strengthening linkages between teachers' and students' EBs?
- To what extent does classroom culture influence the nature of teachers' and students' EBs?
- To what extent do changes in students' EBs influence the EBs of teachers?

Relationships among components of the working model would be expected to vary in cross-cultural studies. For example, the influence of the external culture and external LE (e.g., home environment) would be predictably different when comparing the results of EBs studies of Eastern and Western cultures. Our working model would predict, for example, that student interactions with other students have a strong role in developing student EBs, which is somewhat contrary to the predominant concept that the teacher is the primary influence on developing students' EBs. The model also assumes that any set of EBs developed in classrooms is nested within multiple levels of culture and LEs. Thus, the external LE and external culture (e.g., family/ home environment) can facilitate or hinder student EBs. Thus, EBs developed at the classroom level can be influenced by multiple cultures and multiple LEs. A student's family culture might weaken (or strengthen) EBs considered important by the teacher. Similarly, the working model suggests that teacher EBs can be influenced by interactions with others. These teacher EBs are also nested within and influenced by various levels of culture and LEs. The EBs of a novice teacher, for example, might well be influenced by beliefs of other teachers, particularly veteran teachers. Interactions between levels of the working model, in any combination are possible. Teachers' EBs reflecting the school culture (e.g., beliefs about the tentative nature of knowledge) may be in conflict with those reflecting the teacher's family values (external culture). Thus, in understanding teacher EBs, it would be of interest to study what happens when discrepancies among teacher EBs within a school or perhaps across various cultures are resolved.

Our review of EBs, various cultures and LEs provides numerous examples of the viability of the working model presented here for better understanding and strengthening teachers' and students' EBs. Elements of the model are consistent with selected categories of findings in the Organization for Economic Co-Operation and Development (OECD) (2009) report. Our model is more inclusive of the influences of multiple cultures and multiple LEs on teachers' and students' EBs then the OECD categories. The model also assumes the concept of *reciprocal determinism* among various elements of the model and their interactive nature. The section that follows is a brief literature review on cross-cultural EB studies in Western, Asian, and Middle Eastern cultures. This section is followed by a brief discussion of Turkey as a country in which science education is embedded in several distinct, but transitioning cultures and LEs.

CROSS-CULTURAL STUDIES OF WESTERN, ASIAN, AND MIDDLE EASTERN EPISTEMOLOGICAL BELIEFS

A large number of EBs studies have been completed in western countries, primarily in North America. Furthermore, measures to assess personal epistemology have been largely formulated and validated in the USA (Hofer, 2008) and some have been administered in non-western cultures after simply translating existing measures into non-english languages (Chan, 2008). Such adaptations are made from the fundamental belief that dimensions, stages, and directionality of epistemologies develop similarly across cultures and may be universal.

A considerable body of research on teachers' and students' EBs has been completed to date. The primary focus here is on teacher EBs. As used here *teacher* refers to teacher practitioners and pre-service teachers as well. This line of inquiry includes a considerable number of studies of EBs in different cultures. A number of studies are on going in Europe (Clarebout, Elen, Luyten, & Bamps, 2001) and this line of inquiry is also growing dramatically in Asia, with significant work appearing from Taiwan (Liang & Tsai, 2010), South Korea (So, Lee, Roh, & Lee), the Philippines (Magno, 2011), Hong Kong and China (Chan, 2008), and Singapore (Chai, Khine, & Teo, 2006). Research on EBs is rapidly spreading in the Middle East, with on going studies in Egypt, Saudi Arabia, Lebanon, Israel, and Turkey (Hashweh, 1996; Yilmaz-Tuzun & Topcu, 2008).

Research on interactions between culture and EBs has implications for many social, self, and educational practices. It is by and large assumed that EBs are collectively constructed and, for that reason, culture plays an important role in the formation of EBs (Haerle & Bendixen, 2008). A number of scholars have identified the need for cross-cultural research to investigate similarities and differences in the EBs of students and teachers and the implications of such findings. These studies have identified a number of similarities and differences in EBs across cultures when compared to research in North America (Brownlee et al., 2001). At the present time, our understanding of the cross-cultural perspectives of EBs (beliefs about the nature of knowledge and knowing) seems inadequate.

Currently, cross-cultural research has been mostly completed on dimensions of EBs (Schommer, 1990), such as certainty of knowledge and sources of knowing, and how these perspectives differ in North American and non-Western cultures (Haerle & Bendixen, 2008). Distinctive cultural patterns have been detected in teachers' and students' personal epistemologies. In Eastern cultures, for example, there is an emphasis on collectivism, acceptance of consensus, respect for authority, teacher and student beliefs reflecting the acceptance of external, authoritative sources as experts and the importance of effort in academic achievement (Chan & Elliott, 2004). In contrast, students and teachers in western cultures attribute learning to a combination of external and internal knowledge sources (Brownlee et al., 2001). These beliefs and perspectives are imbedded within larger education cultures that reflect western values of democracy, independent thinking, and individualism (Hofer & Pintrich, 1997).

Using a modified version of Schommer's (1990) 63-item epistemological questionnaire in an Asian cultural context, Chan (2008) completed a series of studies on EBs and the relationship of these beliefs to metacognitive variables in learning and teaching in Hong Kong with teacher education students. Four factors or dimensions documented in these studies resembled but also differed from Schommer's findings. More specifically, Chinese students believed that knowledge is attained through one's personal endeavors and the learning process rather than attained from authority figures or experts. In Chan's study, students did not believe that ability is inborn or fixed and that knowledge is certain and unchanging (similar to findings from Schommer's work). Chan (2008) attributed the differences to Confucian Chinese culture, which places a high value on education, effort, endurance, hard work, and respect for, and obedience to elders and authority figures. In Chinese culture, teachers and educators are considered and respected as sources of knowledge. In an attempt to provide depth to the understanding of EBs and cultural differences between North American students and Chinese students Chan and Elliott (2002) concluded that some of the student EB responses were "at times inconsistent and occasionally contradictory" (p. 404). The results seemed to suggest that conflicting responses may have been due, in part, to domain and contextual differences. In addressing EB measurement issues these researchers concluded, "care needs to be exercised in applying Schommer's questionnaire in another cultural context" (p. 408), a noteworthy caution.

Interestingly, cross-cultural research that has been completed in western contexts (U.S.A. vs. Germany) has produced incongruity as well. In one study, for example, examining the similarities and distinctions among Western cultures, Haerle and Bendixen (2008) studied how German and U.S. elementary school teachers concur and differ in terms of their personal epistemology in the context of German and U.S. elementary school cultures. Analyses of semi-structured interviews with 20 teachers revealed both similarities and differences in teachers' EBs. While the majority of teachers from both countries believed that knowing is uncertain and knowledge has domain-specific qualities, USA teachers seemed to view knowledge as more embedded within their community and their German counterparts discussed more internal knowledge sources (understanding of knowledge as stemming from internal processes of knowing). Haerle and Bendixen (2008) acknowledged that these differences offer insight into the context within which these teachers were trained in terms of educational goals/traditions (e.g., German teachers develop and practice their personal teaching philosophy (Didactic), while teachers in the USA rely more on scientifically based practice). Such differences might stem from a range of possibilities, such as the local cultural context, the culture of the teacher preparation tradition, and/or the community of practitioners.

Compared to the socio-cultural contexts described above, there is a paucity of studies of teacher and student EBs in Middle Eastern contexts. As Karabenick and Moosa (2005) acknowledged, epistemological studies of Middle Eastern populations regarding Western conceptualizations of personal epistemology to Middle Eastern cultures and/or to comparisons between Western and Middle Eastern cultures, especially Muslims, are clearly absent from the literature. Middle Eastern cultures, by and large, reflect characteristics of vertically collective cultures (Triandis, 1995) characterized by authoritarianism, conservatism, emphasis on in-group unity, respect for in-group norms, and the hierarchical directives (Bond & Smith, 1996). In vertical cultures there is submission to higher authority and endorsement of cultural traditions and conventionalism. Vertical collectivism is positively linked to age and religiosity, and negatively linked to education and exposure to diversity (Triandis, 1995). Thus, Trandis (1995), for instance, pointed out the lack of research on individualism–collectivism in Middle Eastern contexts. In such cultures, learners are more apt to recognize omniscient authority-based claims, a prevalent feature of not fully formed EBs (Hofer & Pintrich, 1997). Support for this prophecy so far is based on the evidence obtained from Western and Asian cultural contexts. The section that follows uses Turkey as an example of how historical changes in culture, educational policy, and conceptions of teaching and learning have served to influence EBs systems of teachers and students.

AN EXAMPLE FROM TURKEY

Turkey, a modern Middle Eastern country, offers an interesting context for epistemological research given its Secular, Islamic, Capitalistic economic force, and pluralistic societal structure. Muslims (approximately 99%) (mostly Sunni) and others (mostly Christians and Jews) make up of the population. Turkey is also a multi-ethnic and multi-cultural country. Yet, ethnic groups in Turkey have been subject to the homogenization of state policies, some of which originate from the nationalist Turkish history of 1932, which placed Turks at the center of world civilization. For the past 89 years the Turkish government has pushed for a series of secularist driven educational reforms. Yet, these reforms have been concerned with maintaining equilibrium between the push for Islamic education and the defining of a strong Turkish national identity. The perceptions of Turkish citizens are quite varied and they reflect Turkey's Islamic, Eastern, multinational, long-established Ottoman past and its engineered western, secular, positivist/modern, and national presence. The early years of the republic progressive educational movement had an impact on the Turkish education system. However, this movement did not last long enough to become ingrained in the larger educational culture. Traditional teaching and learning practices have heavily dictated LEs at all levels of education advocating a competitive, rote learning among students leaving little room for creativity, flexibility and freethinking.

For the past few decades, EB studies have found their way into the Turkish educational research agenda. This line of research has to a great extent been focused on students' EBs rather than the EBs of teachers (Kizilgunes et al., 2009; Topcu & Yilmaz-Tuzun, 2009). These studies mainly focus on the issues of gender, metacognition, socioeconomic status, and approaches to learning in constructivist LEs. This research has produced somewhat mixed results. Some studies report statistically significant differences between EBs and other variables, and others report no differences. Most Turkish epistemological studies have been completed with students from grades four through eight in urban settings and with preservice teachers, mostly elementary science teachers[1]. Studies of preservice teachers' have examined their epistemological understandings, views of teaching, learning, and practice, and interactions among these components (Yilmaz-Tuzun & Topcu, 2008). The focus of research in Turkey is not as broad-based as research in North America, Europe, and Asia. Fewer studies have focused solely on the EBs of practicing teachers.

Studies of pre-service science teachers' epistemic beliefs are arguably rare in Turkey. In a few prominent studies researchers were interested in examining the dimensionality of EBs of pre-service teachers (Topcu, 2011; Yilmaz-Tuzun & Topcu, 2008). Yilmaz-Tuzun and Topcu (2008) explored pre-service elementary teachers' EBs from self-efficacy and epistemological worldview perspectives. The results documented an inverse relationship between pre-service as teachers' innate ability beliefs and self-efficacy, confidence, and worldviews. It was found that the less teachers believed in innate ability, the more likely they were to have high self-efficacy in and feel confident about science teaching and were relativist in their epistemological worldview. These teachers also displayed very sophisticated beliefs about innate ability while maintaining naïve beliefs about certain knowledge and simple knowledge. In a subsequent study, Topcu (2011) explored the relationships among elementary pre-service teachers' EBs and moral reasoning. This study produced equivocal results in contrast to research by Yilmaz-Tuzun and Topcu (2008). While innate ability, certain knowledge, and simple knowledge dimensions of epistemic beliefs were present in both studies, an additional factor was observed by Topcu (2011). Yilmaz-Tuzun and Topcu (2008) identified "omniscient authority" as a factor framing EBs, whereas the Topcu (2011) study results attained quick learning as a factor instead of omniscient authority. Topcu (2011) associated this inconsistency with the traditional teacher-centered and heavily standardized testing culture of the Turkish educational system.

In 2004, the Turkish Board of Education initiated a reform-based curriculum focused on student-centered teaching and learning practices. Since then the importance of LEs has been increasingly recognized by Turkish scholars. Thus, constructivist LEs and their influence on students' EBs became a focus of interest to Turkish scholars. Using data collected from sixth grade students, Kizilgunes et al. (2009) proposed a model to explain how EBs, achievement motivation, and meaningful learning are related to student achievement. According to this model it is assumed that EBs directly affect learning approaches and achievement indirectly through their affect on achievement motivation. Ozkal, Tekkaya, Cakiroglu, and Sungur (2009) examined the extent to which eighth grade students' constructivist LE perceptions and scientific EBs differentiated approaches to learning. Students who identified their LE as more constructivist-oriented believed that knowledge is tentative and were very likely to embrace meaningful learning approaches while studying science.

In another constructivist LE study, Ogan-Bekiroglu and Sengul-Turgut (2008) studied the role of constructivist pedagogy on ninth-grade high school students' EBs in a physics course. Using a mixed-method research design (pretest-posttest and semi-structured interviews) the researchers documented that, initially, all students held realist or absolutist (the two least sophisticated) views in the dimensions of certainty, simplicity, source, and justification. Post intervention data showed the constructivist-LE helped students develop more sophisticated EBs. The findings showed that many students moved from the realist to absolutist view, and some even moved into multiple points of view across the various domains. Yet, researchers noted that none of the participants developed an evaluative (most sophisticated) epistemological perspective. Aylin and Geban (2011) observed similar findings in their case-based learning on EBs and attitudes of eleventh grade students toward chemistry. A study by Sahin (2009) of problem-based learning on student understanding of Newtonian concepts and their beliefs about physics revealed that control and experimental group students were no different in their physics-related EBs. Indeed, Sahin (2009) observed deterioration in students' beliefs between pre and post survey administrations. Some studies include concern for multiple perspectives and variables in particular research designs (e.g., studies focused on teacher and student interactions, classroom LEs, and classroom culture).

All of these EB studies in Turkey can be more broadly framed in view of the model shown in Figure 1. For example, sweeping policy changes such as the new reform-based curriculum are filtered through multiple LEs and cultures with the goal of changing the nature of classroom interactions between and among teachers and their students and perhaps teacher and student EBs as well. Indeed, formulation of education policy may have at its foundation the EBs of policy makers themselves. What happens when the pervading epistemology of policy makers is directly at odds with the epistemologies of other key education constituents (e.g., teachers, school administrators, school board members, parents, and even students) in bringing about educational change and reform in science. Alternatively, student and teacher EBs and interactions as they enact a new curriculum reform would predictably have minimal influence on policy makers in the external culture and LE. The model shown in Figure 1 suggests a large number of studies that might be designed to broaden our understanding of EBs, their genesis, their reciprocal influences, and their cultural embeddedness.

DISCUSSION AND IMPLICATIONS

The purpose of this chapter is to present a working model that depicts the interactive relationships between the development of epistemological beliefs of teachers and students, classroom cultures, and classroom LEs. The development and description of a working model was provided as a means of better understanding the complexities of developing and strengthening teachers' and students' EBs. A brief literature review of research studies in different countries was included as well as a more in-depth analysis of different cultures in a single country (Turkey). A brief discussion of our most important findings from this work follows.

It is apparent that the philosophy of a culture and values embedded in a culture have strong implications for the development and strengthening of teacher and student EBs. Countries with different cultures vary in the emphasis placed on traditional vs. constructivist teaching and learning practices (OECD, 2009). Countries that embrace Western European-Socratic philosophy, for example, value questioning knowledge and expect students to evaluate beliefs and to generate personal hypotheses. On the other, hand Confucian Asian philosophy denotes demanding, courteous, abortive, and practical learning, in which learners are expected to grasp defined knowledge. Collectivist cultures such as those in Middle Eastern and Asian countries embrace the cardinal values of reciprocity, obligation, duty, tradition, dependence, obedience to authority, equilibrium, self-development and proper behavior. In contrast, individualistic cultures stress creativity, bravery, self-reliance and individual responsibility as key values. All of these cultural values predictably have a role in shaping both teacher and student EBs.

Turkey was used as an example of various kinds of EBs research embedded in a culture that can be characterized as a bridge between Western and Asian cultures. Most EBs studies in Turkey have been completed with either elementary students and/ or prospective elementary teachers. Studies in Turkey have mostly been completed in major urban cities, especially in Ankara. The demographics and social structures of Turkey are quite diverse. Therefore, the same general findings from an urban environment may not have been observed with research participants from other regions of Turkey. People in rural and remote areas of Turkey are more conservative and traditional than those in urban and Western parts of Turkey. These individuals are also mostly from a low socioeconomic class, and reside in single parent working families (especially fathers). Epistemological beliefs studies of gender are somewhat problematic because of large differences in gender roles in rural and urban areas and eastern and western regions. Our review also shows that research on EBs in Turkey (as in many other countries) is relatively recent.

In some of the studies reviewed, the researcher(s) were interested in examining differences between EBs of elementary students and university level students. In Turkey, student-centered teaching practices have been implemented in science classrooms since 2004. However, most students have not reached the point of questioning or critiquing scientific knowledge handed to them through textbooks, teachers, and scientists. A similar trend has been observed in other cultures, such as Asian and Middle Eastern countries, where there is a tendency not to criticize viewpoints of authority figures. The general culture in Turkey includes norms reflecting the importance of respecting teachers, parents, and other authority figures.

The OECD (2009) report on teaching practices, teachers' beliefs and attitudes shows that Turkish teachers lie somewhere in between constructivist and traditional views of LEs. According to the OECD report,

The preference for a constructivist view is especially pronounced in Austria, Australia, Belgium (Fl.), Denmark, Estonia and Iceland. Differences in the strength of endorsement are small in Brazil, Bulgaria, Italy, Malaysia, Portugal and Spain. Hence teachers in Australia, Korea, northwestern Europe and Scandinavia show a stronger preference for a constructivist view than teachers in Malaysia, South America and southern Europe. Teachers in eastern European countries [including Turkey] lie in between. (p. 94).

By way of summary, the following general findings emerged from our development of a working model for better understanding linkages between teacher and student EBs, LEs and culture, a brief review of the literature on these variables, and a close examination of existing and emerging educational contexts in a single case (Turkey):

- The study of the development and strengthening of teacher (and student) EBs is very complex because of differences in LEs and cultures in which EBs are imbedded.
- The majority of EB studies have been completed with students rather than teachers. Most studies of teachers have been completed with pre-service teachers.
- The great variety of LEs and cultures among different countries makes it difficult to generalize findings from one educational context to another.
- The vast majority of research studies on understanding teacher and student EBs outside of Western cultures are based on work and methodologies of Western scholars. This is somewhat problematic, especially when translating language from one study to the next (e.g., translating a paper and pencil measure from English to Turkish).
- Developing and strengthening teacher and student EBs is greatly influenced by the larger cultural contexts in which EBs are imbedded. For example, the preferred classroom LE in Turkey may include a culture of competition among students (individualism). Whereas the larger Turkish culture is characterized by norms reflecting cooperation, sharing, and helping others (collectivism).
- It is clear from the Turkish case that politics and policy are often at odds with belief systems reflecting scientific thinking. A recent example is the decision by the Turkish government to remove the study of evolution from the standard curriculum. This decision reflects historical cultural conflicts between secular and more deeply rooted Islamic religious philosophies.
- The working model presented (Figure 1) seems to be viable for the development of future teacher and student EB studies and broadening our perspectives of the complexity of such research.

Few studies have been made across different LE levels (classroom, school, and external) and cultures (classroom, school, and external). Large-scale studies seem needed that use more sophisticated research designs and data analysis procedures than those reflected in current studies. Hierarchical Linear Modeling (Raudenbush & Bryk, 2002), for example, might be used to better understand the influence of different levels of LEs (e.g., classroom, school) and culture (e.g., classroom, school) on the development of teacher EBs. Few studies were noted that use experimental or quasi-experimental designs to study the development of teacher EBs and the influence of these beliefs on student outcomes. As well, a continued line of inquiry comparing the development, change and sustainability of both teacher and student EBs seems needed.

NOTE

[1] - In 1997 compulsory education in Turkey became eight years, thus grades 1 through 8 labeled as primary/elementary education. Middle school sections of the schools were abolished. Thus science teachers very also labeled as elementary teachers. Very recently, the Turkish government changed compulsory education from 8 to 12 years. Starting in the Fall of 2012, schools are labeled as elementary (4yrs), middle schools (4yrs), and high schools (4yrs).

REFERENCES

Aylin, C., & Geban, O. (2011). Effectiveness of case-based learning instruction on epistemological beliefs and attitudes toward chemistry. *Journal of Science Education and Technology, 20*(1), 26–32.

Bendixen, L. D., & Feucht, F. C. (Eds.). (2010). *Personal epistemology in the classroom. Theory, research, and educational implications.* New York: Cambridge University Press.

Bond, R., & Smith, P. B. (1996). Culture and conformity: A meta-analysis of studies using Asch's (1952, 1956) line judgment task. *Psychological Bulletin, 119*, 111-137.

Brownlee, J., Purdie, N., & Boulton-Lewis, G. (2001). Changing epistemological beliefs in pre-service teacher education students. *Teaching in Higher Education, 6*, 247–268.

Chai, C. S., Khine, M. S., & Teo, T. (2006). Epistemological beliefs on teaching and learning: A survey among pre-service teachers in Singapore. *Educational Media International, 43*, 285–298.

Chan, K. W. (2008). Epistemological beliefs, learning, and teaching: Honk Kong cultural context. In M. S. Khine (Ed.), *Knowing, knowledge and beliefs: Epistemological studies across diverse cultures* (pp. 267–272). New York: Springer.

Chan, K. W., & Elliot, R. G. (2002). Exploratory study of Hong Kong teacher education students' epistemological beliefs: Cultural perspectives and implications on belief research. *Contemporary Educational Psychology, 27*, 392-414.

Chan, K. W., & Elliot, R. G. (2004). Epistemological beliefs across cultures: Critique and analysis of beliefs structure studies. *Educational Psychology, 24*(2), 123–142.

Chou, S., & Liu, C. (2005). Learning effectiveness in a web-based virtual learning environment: A learner control perspective. *Journal of Computer Assisted Learning, 21*(1), 65-76.

Clarebout, G., Elen, J., Luyten, L., & Bamps, H. (2001). Assessing epistemological beliefs: Schommer's questionnaire revisited. *Educational Research and Evaluation, 7*(1), 53-77.

Dhindsa, H. S. (2005). Cultural LE of upper secondary science students research report. *International Journal of Science Education, 5*, 575-592.

Elen, J., & Clarebout, G. (2001). An invasion in the classroom: Influence of an ill-structured innovation on instructional and epistemological beliefs. *LEs Research, 4*(1), 87–105.

Elen, J., Lowyck, J., & Proost, K. (1996). Design of telematic learning environments: A cognitive mediational view. *Educational Research and Evaluation: An International Journal on Theory and Practice, 2*, 213–230.

Fisher, D. L., & Waldrip, B. G. (1999). Cultural factors of science classroom LEs, teacher student interactions and student outcomes. *Research in Science and Technological Education, 17*(1), 83–96.

Haerle, F. C., & Bendixen, L. D. (2008). Personal epistemology in elementary classrooms: A conceptual comparison of Germany and the United States and a guide for future cross-cultural research. In M. S. Khine (Ed.), *Knowing, knowledge and beliefs: Epistemological studies across diverse cultures* (pp. 151–176). New York: Springer.

Hashweh, M. Z. (1996). Effects of science teachers' epistemological beliefs in teaching. *Journal of Research in Science Teaching, 33*, 47-63.

Hodson, D. (1999). Going beyond cultural pluralism: Science education for sociopolitical action. *Science Education, 83*, 775–796.

Hofer, B. K. (2008). Personal epistemology and culture. In M. S. Khine (Ed.), *Knowing, knowledge and beliefs: Epistemological studies across diverse cultures* (pp. 3–22). New York: Springer.

Hofer, B. K., & Pintrich, P. R. (1997). The development of epistemological theories: Beliefs about knowledge and knowing and their relation to learning. *Review of Educational Research, 67*, 88–140.

Karabenick, S. A., & Moosa, S. (2005). Culture and personal epistemology: U.S. and Middle Eastern students' beliefs about scientific knowledge and knowing. *Social Psychology of Education, 8*, 375–393.

Khine, M. S. (Ed.). (2008). *Knowing, knowledge and beliefs: Epistemological studies across diverse cultures.* New York: Springer.

Kizilgunes, B., Tekkaya, C., & Sungur, S. (2009). Modeling the relations among students' epistemological beliefs, motivation, learning approach, and achievement. *Journal of Educational Research, 102*, 243-255.

Konings, K. D., Brand-Gruwel, S., & van Merrienboer, J. J. G. (2005). Towards more powerful learning environmentss through combining the perspectives of designers, teachers, and students. *British Journal of Educational Psychology, 75*, 645–660.

Levy, J., den Brok, P., Wubbels, T., & Brekelmans, M. (2003). Students' perceptions of interpersonal aspects of the learning environment. *Learning Environments Research, 6*(1), 5–37.

Liang, J. C., & Tsai, C.-C. (2010). Relational analysis of college science- major students' epistemological beliefs toward science and conceptions of learning science. *International Journal of Science Education, 32*, 2273–2289.

Magno, C. (2011). Exploring the relationship between epistemological beliefs and self-determination. *The International Journal of Research and Review, 7*(1), 1-23.

Ogan-Bekiroglu, F., & Sengul-Turgut, G. (2008). *Does constructivist teaching help students move their epistemological beliefs in physics through upper levels?* From the Proceeding of Conference of Asian Science Education. Retrieved from ERIC database. (ED500844)

Organization For Economic Co-Operation And Development. (2009). *Teaching practices, teachers' beliefs and attitudes. Creating effective teaching and learning environments: First results from TALIS.* Retrieved October 14, 2012, from http://www.oecd.org/education/preschoolandschool/43023606.pdf

Ozkal, K., Tekkaya, C., Cakiroglu, J., & Sungur, S. (2009). A conceptual model of relationships among constructivist LE perceptions, epistemological beliefs, and learning approaches. *Learning and Individual Differences, 19*(1), 71-79.

Perry, W. G. (1970). *Forms of intellectual and ethical development in the college years:* A scheme. New York: Holt, Rinehart, and Winston.

Raudenbush, S. W., & Bryk, A. S. (2002). Hierarchical linear models applications and data analysis methods. Thousands Oak, CA: Sage Publications.

Sahin, M. (2009). Correlations of students' grades, expectations, epistemological beliefs and demographics in a problem-based introductory course. *International Journal of Environmental Science Education, 4*, 169–184.

Schommer, M. (1990). Effects of beliefs about the nature of knowledge on comprehension. *Journal of Educational Psychology, 82*, 498–504.

Schommer, M. (1994). Synthesising epistemological beliefs research: Tentative understandings and provocative confusions. *Educational Psychology Review, 6*, 293-319.

So, H.-J., Lee, J.-Y., Roh, S.-Z., & Lee, S.-K., (2010). Examining epistemological beliefs of pre-service teachers in Korea. *The Asia-Pacific Education Researcher, 19*(1), 79-97.

Tolhurst, D. (2007). The influence of LEs on students' epistemological beliefs and learning outcomes. *Teaching in Higher Education, 12*, 219-233.

Topcu, M. S. (2011). Turkish elementary student teachers' epistemological beliefs and moral reasoning. *European Journal of Teacher Education, 34*(1), 99–125.

Topcu, M. S., & Yılmaz-Tuzun, O. (2009). Elementary students' metacognition and epistemological beliefs considering science achievement, gender and socioeconomic status. *Elementary Education Online, 8*, 676-693.

Triandis, H.C. (1995). *Individualism and collectivism*. Boulder, CO: Westview.

Tsai, C. C. (2003). Taiwanese science students' and teachers' perceptions of the laboratory learning environments: Exploring epistemological gaps. *International Journal of Science Education, 25*, 847-860.

Yilmaz-Tuzun, O., & Topcu, M. S. (2008). Relationships among preservice science teachers' epistemological beliefs, epistemological worldviews, and self-efficacy beliefs. *International Journal of Science Education, 30*(1), 65–85.

AFFLIATIONS

Kadir Demir
Department of Middle and Secondary Education
College of Education
Georgia State University

Chad D. Ellett
CDE Research Associates, Inc.
Watkinsville, GA

P. SEAN SMITH, ADRIENNE A. SMITH & ERIC R. BANILOWER

SITUATING BELIEFS IN THE THEORY OF PLANNED BEHAVIOR

The Development of the Teacher Beliefs about Effective Science Teaching Questionnaire

This material is based upon work supported by the National Science Foundation under Grant Nos. DUE-0335328 and DUE-0928177. Any opinions, findings, and conclusions or recommendations expressed in this material are those of the authors and do not necessarily reflect the views of the National Science Foundation.

INTRODUCTION

In this chapter, we describe the development of the Teacher Beliefs about Effective Science Teaching (TBEST) survey, a new instrument designed to fill a gap in the collection of extant beliefs measures. The survey was originally designed for a study investigating relationships among science teacher attributes, classroom practice, and student learning. The TBEST draws upon Ajzen's Theory of Planned Behavior (2012, 1985), which posits a model for how beliefs influence actions. The measure was also informed by cognitive science research; specifically, what this research suggests about instruction aimed at conceptual change. The chapter begins by discussing each of these contexts briefly and situating the TBEST in the landscape of beliefs measures. We then describe the process used to develop survey items, which included item writing, cognitive interviews with teachers, and multiple rounds of piloting and revision. We discuss the analyses of data from several pilot studies—which culminated in the identification of three distinct factors—and future directions for research utilizing the TBEST. Although our work revolved around the development of a particular instrument, the lessons learned and principles employed apply to the development of any beliefs measure. We describe these principles in the chapter's conclusion.

THEORETICAL BACKGROUND

The TBEST was developed in the context of a study that examined impacts of science teacher professional development on student learning. Specifically, the study investigated relationships among professional development, teacher attributes, classroom practice, and student learning. A simplified theory of action of professional development is shown in Figure 1.

R. Evans et al., (Eds.), The Role of Science Teachers' Beliefs in International Classrooms, 81–102.
© 2014 Sense Publishers. All rights reserved.

Despite a thorough search of the literature, we were unable to find measures of pedagogical content knowledge[1] (PCK) and teacher beliefs that aligned well with the goals of the study. As a result, only two independent variables were included in the analysis: instructional time on the target science topic and a measure of teacher content knowledge. Together, these factors accounted for only a very small proportion of the variance in student scores. The findings motivated us to develop a new instrument to measure teacher beliefs about effective science instruction. The purpose of the TBEST is to allow researchers to investigate teacher beliefs as both a dependent variable (e.g., an outcome of professional development) and as an independent variable (e.g., a predictor of classroom practice and student learning). The instrument is currently being used in several such studies.

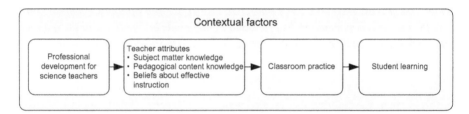

Figure 1. Simplified Theory of Action of Professional Development

The TBEST is based on the Theory of Planned Behavior, a particularly prominent and comprehensive framework in social psychology for thinking about human behavior (Ajzen, 2012; Ajzen, 1985).[2] The theory holds that three types of beliefs indirectly influence behavior: behavioral beliefs, normative beliefs, and control beliefs. Beliefs regarding the expected outcome of a behavior, along with subjective values about the outcome, influence an individual's attitude toward the behavior. To illustrate, a science teacher may believe that allowing students to engage in hands-on activities will result in a chaotic classroom. If the teacher attaches a negative value to a disorderly classroom, then the teacher's overall attitude toward hands-on activities may be negative. Furthermore, the theory holds that attitude toward a behavior is the sum of all behavioral beliefs and subjective values. The same teacher may also believe that hands-on activities will lead to improved student learning. If this belief and the associated value of the outcome are strong enough, they may outweigh concerns about classroom order, and the teacher may have an overall positive attitude toward hands-on activities.

Normative beliefs, or what we believe influential others will think about us if we exhibit the behavior, combined with our motivation to comply, form a subjective norm. Continuing with the example above, if the teacher works in a school or district where administrators advocate for hands-on science instruction, and if the teacher is motivated to comply with the administrators, then the teacher may have a positive subjective norm toward this behavior. If, on the other hand, teachers in the same grade level have taken a stance against hands-on instruction, the teacher's subjective norm may be less positive or perhaps even negative, depending on how motivated the teacher is to comply with her peers.

Finally, an individual's ability to engage in a behavior has to do with factors both internal and external to the individual. The individual's perception of the presence of these factors, along with the perceived power of each factor, constitute control beliefs, which shape the individual's perceived behavioral control. If a school or district does not have the materials needed in order for a teacher to use hands-on activities (an external factor), the teacher may have low perceived behavioral control.

Attitude toward a behavior, subjective norm, and perceived behavioral control influence each other and together shape an individual's intention toward the behavior, which, combined with actual behavioral control,[3] predicts the likelihood of the behavior. Figure 2 represents these relationships graphically (Ajzen, 2008).

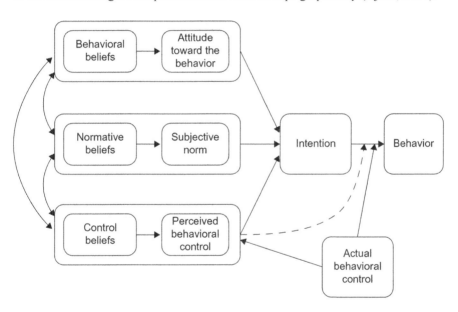

The Theory of Planned Behavior is useful for situating our work because it elucidates the relationship between teacher behavioral beliefs—which are the focus of the TBEST—and actual behavior. (See Figure 2.) The theory is also useful for explaining why behavior is sometimes not manifested, even when behavioral beliefs are favorable. Using the illustration above, a teacher may have a positive attitude towards hands-on instructional strategies and a positive subjective norm, but perceive little control over the behavior because her school lacks the necessary materials. This teacher may sincerely "talk the talk" about hands-on strategies, but have little intention of "walking the walk." Similarly, the Theory of Planned Behavior can explain why science kits sometimes sit unused on teachers' shelves and in school district warehouses. Despite having adequate resources, teachers may believe that the kits will not benefit students or may feel pressure from peers to adhere to a pacing guide.

The Landscape of Extant Teacher Beliefs Questionnaires

There is no scarcity of instruments that purport to capture teacher beliefs related to science and science teaching. To explain why we developed yet another survey, it is necessary to describe the landscape of beliefs measures. First, however, we draw an important distinction between beliefs about science teaching behaviors, which are the focus of the TBEST, and beliefs about science. As an example of the latter type, the Views about the Nature of Science (V-NOS) questionnaire (Lederman, Abd El-Khalik, Bell, & Schwartz, 2002) is a well-known, widely used beliefs measure. As the name suggests, items ask respondents to describe their views toward science as a discipline or way of knowing.[4] Two sample items ask:

- Is there a difference between scientific knowledge and opinion? Give an example to illustrate your answer.
- How are science and art similar? How are they different? (p. 505)

A second example, the Thinking about Science Survey Instrument (TSSI) (Cobern & Loving, 2002), consists of 60 Likert-scale items. Like the V-NOS, it asks about views toward science, not science teaching. For instance, respondents are asked to rate their agreement with the statement, "Scientific explanations tend to spoil the beauty of nature." (p. 1024)

In contrast to the V-NOS and the TSSI, other questionnaires focus on beliefs about science teaching behaviors, a number of which address science teacher self-efficacy. Self-efficacy questionnaire items typically align with the two dimensions described by Bandura (1997): personal self-efficacy and outcome expectancy. For example, the Science Teaching Efficacy Beliefs Instrument (Riggs & Enochs, 1990), or STEBI, consists of 23 statements about science teaching. The items form two scales, one for each of Bandura's dimensions. Teachers respond on a Likert-type scale ranging from strongly disagree to strongly agree. For instance:

- I find it difficult to explain to students why science experiments work. (personal self-efficacy)
- Students' achievement in science is directly related to their teacher's effectiveness in science teaching. (outcome expectancy) (p. 635)

In contrast to the STEBI, the Teaching Science as Inquiry (TSI) instrument (Smolleck, Zembal-Saul, & Yoder, 2006) targets teacher self-efficacy for a particular kind of science teaching: inquiry-oriented instruction. The questionnaire includes 69 statements, divided between the personal self-efficacy and outcome expectancy dimensions. Sample items addressing these dimensions include:

- Personal self-efficacy
  - I possess the ability to allow students to devise their own problems to investigate.
  - I will be able to play the primary role in guiding the identification of scientific questions.
- Outcome expectancy
  - My students will engage in questions I have provided them.
  - My students will construct explanations from evidence using a framework I have provided. (p. 293–294).

Self-efficacy beliefs can be thought of in terms of the Theory of Planned Behavior. Personal self-efficacy aligns with Ajzen's notion of perceived behavioral control, and outcome expectancy is similar to behavioral beliefs. According to the Theory of Planned Behavior, teacher beliefs about the context in which they work, along with their sense of self-efficacy, affect their perceived behavior control. As mentioned above, teachers in resource-poor settings may feel that the context constrains their use of instructional strategies. The Context Belief About Science Teaching (CBAST) instrument (Lumpe, Haney, & Czerniak, 2000) taps this aspect of teacher beliefs. The instrument presents 26 contextual factors and asks respondents to rate: (1) the extent to which each factor would enable effective science teaching; and (2) the likelihood that each factor would occur. Sample factors include planning time, support from administrators, and involvement of scientists.

The Beliefs About Reformed Science Teaching and Learning (BARSTL) instrument (Sampson & Benton, 2006) consists of 32 statements in four subscales, each subscale aligned with a different aspect of science teaching and learning. As the name implies, the content for the instrument is drawn from a framework of reform-oriented science teaching (National Research Council, 1996). Respondents indicate their agreement on a four-point response scale from strongly disagree to strongly agree. In general, the statements have an implied "if, then" structure; i.e., if the teacher/curriculum does X, students will learn. From a Theory-of-Planned-Behavior perspective, the instrument taps respondents' behavioral beliefs. The four subscales and a sample statement from each are shown in Table 1 (p. 41–42).

Situating the TBEST in the Landscape

Within the Theory of Planned Behavior, the TBEST clearly targets the category of behavioral beliefs. As described in detail below, the questionnaire asks teachers to indicate their agreement that particular behaviors will have a particular outcome; specifically, student learning of science concepts. The Theory of Planned Behavior highlights the fact that many factors besides beliefs influence behavior. Even when a teacher believes that a behavior will lead to positive outcomes, social pressures may come into play (subjective norms), and resources may affect teachers' perceived behavioral control. In developing the TBEST, we chose to focus on behavioral beliefs despite all of the factors that mediate their relationship with actual behavior

Table 1. Subscales and Sample Statements from the BARSTL

Subscale	Sample Statement
How people learn about science	Students learn the most when they are able to test, discuss, and debate many possible answers during activities that involve social interaction.
Lesson design and implementation	During a lesson, students should explore and conduct their own experiments with hands-on materials before the teacher discusses any scientific concepts with them.
Teachers and the learning environment	Science teachers should primarily act as a resource person; working to support and enhance student investigations rather than explaining how things work.
The science curriculum	A good science curriculum should focus on only a few scientific concepts a year, but in great detail.

(in contrast to the STEBI and CBAST, which include items aimed at understanding other emotional or contextual factors that may affect behavior). Of the three types of beliefs, behavioral beliefs are, in our experience, the most likely to be targeted by science teacher professional development and are perhaps the most malleable.

The content of the questionnaire items is situated broadly within a theory of conceptual change (Posner, Strike, Hewson, & Gertzog, 1982). That is, although science instruction may target various kinds of goals—among them, improving students' attitudes toward science or learning facts/definitions/algorithms—we were interested in instruction aimed at helping students build understanding of concepts. Some of the extant questionnaires have a similar focus; e.g., the TSI and the BARSTL described above. However, these instruments are not explicitly situated in research on learning.

We chose to base the TBEST on an instructional model (Banilower, Cohen, Pasley, & Weiss, 2008) informed by the research on learning summarized in the National Research Council's volumes *How People Learn: Brain, Mind, Experience, and School* (2000) and *How Students Learn: Science in the Classroom* (2005). The model does not prescribe specific pedagogies, as it is quite possible for a pedagogy to be used effectively or ineffectively. Rather, it describes five elements of instruction, each of which could be accomplished using a variety of pedagogies. The elements are: motivating learners, surfacing their prior knowledge about the idea, engaging them with phenomena that provide evidence for the idea, using evidence to make and critique claims, and making sense of the targeted idea. These elements, as described by Banilower et al. (2010), are summarized briefly below.

Motivation. Effective instruction ensures that students are motivated, either intrinsically (e.g., by a discrepant event, providing a real-world context, or a problem to solve) or extrinsically (e.g., by grades or tests).

Surfacing prior knowledge. Students come to school with ideas and beliefs—gleaned from books, television, movies, and real-life experiences—which may facilitate or impede learning. Surfacing prior knowledge is important so teachers can plan and adjust instruction accordingly. Students also benefit from being aware of their prior thinking, as it provides them with ideas to test and facilitates metacognition.

Engaging with phenomena. Consistent with the nature of science, effective science instruction should intellectually engage learners with phenomena that provide evidence for the target idea. Although hands-on experiences may be necessary for students to learn some ideas, particularly ones students have strongly held naive conceptions about, classroom experiences do not always have to be hands-on in order to engage students. Students can be intellectually engaged with an interactive lecture that encourages them to consider examples of the idea from their everyday lives. If hands-on experiences are used, they should reliably provide evidence for the target idea. An experiment that does not adequately control variables, is prone to large measurement error, or is otherwise likely to yield flawed data, may result in students drawing conclusions that are supported by their data, but are inconsistent with the accepted scientific view.

Using evidence. Science is an evidence-based discipline, and having learners use evidence to make and critique claims models the practice of science and facilitates learning. Cognitively, using evidence from the phenomena they have engaged with helps learners make the connections between the instructional activities and the learning goals. In addition, the more evidence for an idea learners engage with, the more likely they will be to reconsider and reconcile their initial ideas with the more scientifically accepted ideas.

Making sense. Effective science instruction requires explicit opportunities for students to make sense of the ideas they have explored. Sense making can occur in a variety of ways. Students may be encouraged to make connections between what they did in the lesson and what they were intended to learn so that they see a purpose to their activities. Students may also be asked to reflect on their initial ideas, becoming aware of how their thinking may have changed over the course of the lesson or unit. Another aspect of sensemaking involves helping students connect the target ideas to what they have learned previously, organizing their new knowledge in a larger cognitive framework. Finally, students may be given opportunities to apply the concepts to new contexts, helping to reinforce their understanding of and increase their facility with the ideas.

DEVELOPING SURVEY ITEMS

Item development for the TBEST followed a rigorous process established for writing and refining science assessment questions. The process began by specifying the content domain—the elements of instruction described above. Within an assessment development framework, each of the five elements of effective science instruction corresponds to an idea. After identifying the content domain, the next step was to unpack each idea, or element, into "sub-ideas." An example is shown in Table 2. This process was surprisingly complex. A group of science education researchers met regularly over a period of several weeks to reach consensus on the sub-ideas. Discussions of the elements of effective instruction surfaced unexamined assumptions and revealed disagreements among researchers who previously thought their beliefs were closely aligned. For instance, one of the ideas related to student motivation is, "Learning is enhanced when students can recognize a purpose of what they are doing in a lesson." In conversation, some researchers expressed the view that as long as students had any sense of purpose, the criterion had been met. Others thought that the students' perceived purpose should align closely with the teacher's instructional purpose. As an example, students might believe that the purpose of a lesson was to build the best water balloon launcher, while the teacher's purpose was for students to engage with ideas about projectile motion. Because the researchers could not reach agreement, no items were developed for this idea.

The next step in the development process was to write survey statements related to each sub-idea. Like the unpacking conversations, this process generated spirited discussions among researchers about what effective science instruction looks like. A common refrain was, "It's not practical to incorporate all of the elements all of the time." That is, if teachers always include all of the elements in their instruction, they would not be able to address all of the content they are charged with teaching. We also had lengthy discussions about whether all of the elements are necessary for every science concept. Some researchers argued that cognitive science literature has been shaped substantially by studies in the physical sciences, in which students tend to have deeply held misconceptions. In these areas, research suggests that students need to experience all of the elements in order to form concepts that align with current scientific thinking. Guidance from research is less clear when students do not have strongly held misconceptions. For instance, our assessment development work suggests that students often do not have strongly held misconceptions about Earth's tectonic plates. Students often have misinformation, but they have not formed incorrect ideas through daily interactions with plates, as is often the case in the physical sciences. Ultimately, we decided that it was not feasible for the TBEST to be concept specific and would instead be consistent with cognitive science findings, even if that research was not completely representative of all science topics.

We also decided to ask teachers to ignore practical constraints of the classroom. Each time we tried to account for these constraints, the questions became more about what teachers actually do in the classroom than about their beliefs. We constructed the following preamble in the instructions for the questionnaire, the purpose of which was to help teachers focus on their behavioral beliefs and set aside their normative beliefs and control beliefs (Ajzen, 2012).

We recognize that teachers have to make many trade-offs when they are responsible for teaching many standards in one year. Teachers may not be able

Table 2. Sample Unpacking of an Idea in the Content Domain

Element of instruction: *Instruction should engage the learner with phenomena that provide evidence for the targeted idea.*

Sub-ideas

Learning is enhanced when students have opportunities to engage with phenomena that provide data that are *relevant* to the targeted content.

Learning is enhanced when students have opportunities to engage with phenomena that are *appropriate* in terms of the students' life experiences.

Learning is enhanced when students have opportunities to engage with data that are *sufficiently precise* to form the science concept.

Learning is enhanced when students have opportunities to engage with phenomena for which *students can collect their own data.*

to emphasize the instructional strategies they believe are effective and still cover the entire curriculum. When you respond to the statements below, we ask that you put those trade-offs aside. Imagine that you have no constraints, including state/district standards, available time and resources, and feasibility. We want to know what you think effective instruction looks like, without all the constraints that limit what you can do in the classroom.

Discussions about response options also occupied many development meetings. For the initial version of the questionnaire, we eventually settled on two response-option formats: agreement and importance. The stem for the first asked teachers whether they agreed with each statement; the second asked teachers about the importance of students experiencing what was described in the statement. To the extent possible, we wrote parallel versions of statements for each response-option format, as we thought it was important to test both and learn which one produced greater variation in teacher responses.[5] For example, a statement about eliciting students' prior knowledge resulted in the following two items:

Practical constraints aside, do you *agree* that doing what is described in each statement would help most students learn science?

Teachers should be aware of their students' prior knowledge of a science topic before the lesson begins. (four response options ranging from "strongly disagree" to "strongly agree")

Practical constraints aside, how *important* is each of the following for helping students learn science?

Teachers are aware of their students' prior knowledge of a science topic before the lesson begins. (seven response options ranging from "very important that this does not happen" to "very important that this does happen")

Using this approach, we wrote multiple statements for each sub-idea. We then conducted cognitive interviews (Desimone & Le Floch, 2004) about the items by telephone with 17 middle grades science teachers throughout the United States. The purpose of these interviews was to ensure that teachers interpreted the statements as we intended. One issue illustrates the importance of these interviews in the development process. Many of the science education reform documents (e.g., National Research Council, 1996; American Association for the Advancement of Science, 1993) make frequent use of the term "phenomena" to represent naturally occurring events with which students should engage. Many of the original TBEST items used this term as well. Teachers, however, largely interpreted "phenomena" quite differently, thinking instead of *supernatural* events; not at all what we had in mind. We felt compelled to remove the term from the questionnaire and use alternatives. The cognitive interviews suggested other edits but none as pervasive as this one. The months-long development process yielded just over 100 questionnaire statements, approximately 50 for each response-option format.

PILOTING AND DATA ANALYSIS

In this section, we describe the piloting process and the analyses that led to the final version of the questionnaire, which consists of 21 items with a six-point agreement response-option format, organized in three factors. Each pilot contributed an additional piece of evidence for the validity and reliability of the TBEST.

As described above, researchers composed over 100 items intended to conceptually align with the five elements of effective science instruction. Approximately 950 middle grades science teachers responded to the first pilot of the items, which was conducted online. A number of important and related findings emerged from the data. First, the four-point agreement response-option formats did not generate sufficient variation in teacher responses. (Several had no variation in responses and were eliminated from the survey.) Second, the data suggested that some respondents did not answer the questions thoughtfully. For instance, some individuals gave the same response to adjacent items that had opposite meanings. Our hypothesis was that the lack of thoughtfulness was due to the length of the questionnaire.

Based on the results, we chose the importance response-option scale and 23 items for the second phase of piloting, also conducted online. The items were chosen based on coverage of the content domain and variation in responses. Middle grades science teachers were recruited for participation, and an exploratory factor analyses (EFA) was conducted on the resulting sample of just under 250 respondents. The EFA was run using an oblique rotation,[6] which allowed any underlying factors to correlate. The analysis suggested five factors, which, based on the items, were labeled: (1) the importance of situating learning; (2) the importance of using evidence in sense making; (3) the importance of connecting new learning and prior learning; (4) the importance of using activities to confirm concepts that have already been taught (which we refer to as confirmatory instruction); and (5) the importance of hands-on instruction. However, some of the factors were highly correlated (e.g., the correlation between situating learning and confirmatory instruction was –0.57), causing concern about whether the factors were indeed distinct dimensions.

In order to assess the robustness of the five-factor structure, a third pilot was conducted. At this point, we addressed a disconcerting feature of the survey. Although the importance response-option format produced sufficient variation in responses, it seemed a force fit for many of the statements, requiring respondents to mentally alter the item or the response options to create alignment. Rather than continue with this response-option format, we returned to the agreement format but expanded it to six points, rewording the items to make them appropriate for the response options. The result was much better alignment between the items and the response options.

Approximately 250 middle grades science teachers responded to the new version of the questionnaire. Using the five-factor solution suggested by the EFA, a confirmatory factor analysis (CFA) using Mplus version 5.2 was applied. However, the CFA results did not support the five-factor solution, and follow-up analyses suggested a three-factor solution was more appropriate. The three factors were conceptually coherent and were labeled: (1) Learning-theory-aligned science instruction; (2) Confirmatory science instruction; and (3) All hands-on all the time.

Next, we investigated the psychometric soundness of the survey's underlying structure across administration modes (paper versus online) and grade levels (K–12). In the first of these studies, just over 600 teachers were randomly assigned to receive either an online or paper version of the instrument. The previous pilots had been exclusively online; however, we anticipated that other researchers might prefer a paper-and-pencil version. Therefore, it seemed important to establish that similar results would be obtained regardless of administration mode. We decided to conduct an EFA on data from the paper version followed by a CFA on data from the web version. The same three-factor solution fit for both modes of administration, and there were no statistically significant differences in factor composite means, suggesting that the instrument produces similar scores regardless of whether it is administered on paper or online.

We were also interested in the robustness across grade levels, anticipating that researchers might want to use the TBEST in studies of elementary, middle, or high school science teaching. A final study was designed in which we administered the TBEST to a total of 900 elementary, middle, and high school teachers. To test whether the factor structure was the same across grade levels, a multiple-group CFA procedure was followed, again using Mplus version 5.2. This procedure involves conducting an initial CFA for each grade range separately, followed by a multiple-group CFA.

The individual grade-range CFAs pointed to the previously identified three-factor structure. Modification indices provided by the software identified two items that did not fit well with the three-factor structure, and these items were subsequently dropped from the survey. The adequacy of model fit for each grade range was assessed; typically researchers examine a number of indices, using a somewhat holistic approach to judging model fit (Schumacker & Lomax, 1996). For this analysis, we used the fit indices available in the software package: the Chi-Square Goodness of Fit test, the CFI, the TLI, and the RMSEA.[7] A significant Chi-Square test indicates that the model is not an adequate fit of the data; however, this test is very sensitive to sample size, and with the large samples used in our study, is not a good measure of fit (Tabachnik & Fidell, 2007). The research community has debated the best criteria for judging fit on each of the remaining indices. We elected to use the traditional criteria, where a good fit is defined as: CFI > 0.9, TLI > 0.9, and RMSEA < 0.08 (Browne & Cudeck, 1993). As can be seen in Table 3, the fit indices provide evidence of the appropriateness of the three-factor solution for each grade range.

Table 3. CFA Model Fit Indices by Grade Range Model

	Chi-Square Goodness of Fit test	*CFI*	*TLI*	*RMSEA*
Criteria for good-fit:	*Not statistically significant*	*> 0.9*	*> 0.9*	*< 0.08*
Elementary	$\chi 2(71, N = 332) = 207.319, p < .01$	0.929	0.940	0.076
Middle	$\chi 2(64, N = 262) = 149.210, p < .01$	0.940	0.949	0.071
High	$\chi 2(75, N = 372) = 257.461, p < .01$	0.912	0.930	0.081

Because not all response options were chosen by respondents in each grade range, it was not possible to run the multi-group CFA. However, other results provide support for the same three-factor model for each grade range. First, the factors were not highly correlated with each other, suggesting distinct constructs. (See Table 4.) Furthermore, the reliabilities (Cronbach's alpha) of the composites for each grade range are above 0.70. (See Table 5.) These findings were consistent across all grade ranges.

Table 4. Correlations Among Factors[†]

	Learning-Theory-Aligned Science Instruction	*Confirmatory Science Instruction*	*All Hands-on All the Time*
Learning-Theory-Aligned Science Instruction	1.00		
Confirmatory Science Instruction	-0.18	1.00	
All Hands-on All the Time	-0.07	0.45	1.00

[†]Factor correlations were similar across grade ranges

Table 5. Cronbach's Alpha Reliability Coefficients by Grade Range Taught

	Grade Range			
	Overall *(N = 966)*	*Elementary* *(N = 332)*	*Middle* *(N = 262)*	*High* *(N = 372)*
Learning-Theory-Aligned Science Instruction	0.713	0.766	0.739	0.761
Confirmatory Science Instruction	0.771	0.758	0.775	0.784
All Hands-on All the Time	0.758	0.794	0.747	0.732

To summarize, the resulting questionnaire contains 21 items using a six-point agreement response scale. The items fall into three factors: (1) Learning-theory-aligned science instruction; (2) Confirmatory science instruction; and (3) All hands-on all the time. Statistical findings support the psychometric structure of the survey across different modes of administration and across teachers of various grade ranges. Table 6 shows the items organized by factor; a copy of the instrument is available in the appendix.

Table 6. Questionnaire Factors and Associated Items

Factor 1: Learning-Theory-Aligned Science Instruction (11 items)

Q3: Students should rely on evidence from classroom activities, labs, or observations to form conclusions about the science concept they are studying.

Q6: Teachers should provide students with opportunities to connect the science they learn in the classroom to what they experience outside of the classroom.

Q7: Teachers should ask students to support their conclusions about a science concept with evidence.

Q9: At the beginning of instruction on a science concept, students should have the opportunity to consider what they already know about the concept.

Q11: Teachers should provide students with opportunities to apply the concepts they have learned in new or different contexts.

Q12: Students should use evidence to evaluate claims about a science concept made by other students.

Q14: At the beginning of lessons, teachers should 'hook' students with stories, video clips, demonstrations or other concrete events/activities in order to focus student attention.

Q15: Students' ideas about a science concept should be deliberately brought to the surface prior to a lesson or unit so that students are aware of their own thinking.

Q17: Students should have opportunities to connect the concept they are studying to other concepts.

(Continued)

Table 6. Continued

Q18: Students should consider evidence that relates to the science concept they are studying.

Q21: Students should consider evidence for the concept they are studying, even if they do not do a hands-on or laboratory activity related to the concept.

Factor 2: Confirmatory Science Instruction (7 items)

Q1: At the beginning of instruction on a science concept, students should be provided with definitions for new scientific vocabulary that will be used.

Q2: Hands-on activities and/or laboratory activities should be used primarily to reinforce a science concept that the students have already learned.

Q5: Teachers should explain a concept to students before having them consider evidence that relates to the concept.

Q10: Students should do hands-on activities after they have learned the related science concepts.

Q16: Teachers should provide students with the outcome of an activity in advance so students know they are on the right track as they do the activity.

Q19: When students do a hands-on activity and the data don't come out right, teachers should tell students what they should have found.

Q20: Students should know what the results of an experiment are supposed to be before they carry it out.

Factor 3: All Hands-on All the Time (3 items)

Q4: Teachers should have students do hands-on activities, even if the data they collect are not closely related to the concept they are studying.

Q8: Students should do hands-on or laboratory activities, even if they do not have opportunities to reflect on what they learned by doing the activities.

Q13: Teachers should have students do interesting hands-on activities, even if the activities do not relate closely to the concept being studied.

The TBEST, like any social science measure, is vulnerable to social desirability bias—the tendency for a respondent to provide answers that are consistent with social norms. In this questionnaire, if a researcher attended to the learning-theory-aligned composite score exclusively, a high score could indicate a teacher whose beliefs about effective science instruction were truly aligned with cognitive learning theory or a teacher who answered affirmatively to the items because she thought those responses were the most socially acceptable. To distinguish between these possibilities, a researcher would benefit from considering the profile of three composite scores provided by the TBEST. On the TBEST, a teacher whose beliefs about effective science instruction were aligned with cognitive learning theory would score high on the learning-theory-aligned-instruction factor and score low on the confirmatory science instruction and hands-on all the time factors. Considering the profile of scores on all three factors allows for a more nuanced understanding of a teacher's beliefs.

SUGGESTIONS FOR FUTURE RESEARCH

Teacher beliefs and their role in shaping what teachers do in the classroom have long been, and continue to be, a subject of interest in education. The Theory of Planned Behavior in general and the TBEST specifically open a number of interesting avenues for future research, including testing the applicability of the theory in science education. For example, different kinds of beliefs instruments, along with measures of classroom practice, could be used to examine the role that behavioral, normative, and control beliefs play in instructional decision making.

Consistent with the importance of looking at a respondent's profile of scores, we conducted a cluster analysis on the composite scores from the TBEST pilot data and identified four main groups of respondents. Based on their patterns of responses, teachers tended to believe that:

1. Learning-theory-based instruction is most appropriate; or
2. Hands-on instruction is always the right way to go; or
3. Confirmatory instruction works best; or
4. Instruction should include a little of each of these approaches (all things in moderation).

It would be interesting to study the science instruction of people in each of these groups to examine the extent to which group membership predicts classroom practice. Such a study would also provide an opportunity to examine which factors mediate the relationships between beliefs and practices, such as the nature of instructional materials available.

The TBEST could also be used to examine the malleability of teachers' behavioral beliefs about effective science instruction, and, equally important, the ways in which those beliefs can be changed. If the TBEST was found to be valid for measuring pre-service teacher beliefs, it could be used to examine the impacts of different pre-service programs. It could also be used to examine the effect of different professional development experiences on teacher beliefs, such as new teacher mentoring. It might also be used to examine the belief structures of other stakeholders (e.g., administrators, parents) and how those beliefs evolve over time.

These examples are not meant to be exhaustive. Rather, they illustrate the rich variety of ways the TBEST could be used. We are confident that researchers interested in science teachers' beliefs, science teaching, and efforts aimed at improving science education more broadly will identify other appropriate ways to use the TBEST.

CONCLUSION

The purpose of this chapter was to describe the development of a science teacher beliefs measure in order to illustrate several overarching design principles. To conclude the chapter, we identify and briefly discuss these principles, with the aim of providing guidance for other developers and for researchers using belief instruments.

All beliefs measures are designed for a specific purpose, which shapes both the instrument and how others will be able use it. As described in the chapter, the

TBEST was developed to serve as a measure of both a dependent variable (studying the effects of professional development) and an independent variable (studying how beliefs influence instructional practice). Researchers should be aware of an instrument's original purpose in deciding how well it aligns with aims of their own work.

Developers should carefully define the domain of beliefs in which they are working before they begin writing survey items. In the chapter, we used the analogy of a landscape to describe the space occupied by our measure. We argued that the TBEST fills a gap in the landscape by measuring teacher beliefs about effective science teaching, as informed by cognitive science research. A careful description of an instrument's place in the landscape helps developers stay focused on their purpose and helps researchers judge the alignment of the instrument with their research goals.

Developers should be clear about their theory of action. That is, they should explain how the beliefs they are focusing on are shaped and how those beliefs influence action. Like carefully defining the domain of beliefs, having a well-specified theoretical framework keeps developers focused and helps users judge the appropriateness of the instrument for their purposes. The TBEST was designed to measure behavioral beliefs about science instruction, recognizing that other kinds of beliefs come into play, and that other factors mediate the relationship between beliefs and actions. Consequently, disconnects between beliefs and behaviors are common. By specifying the theory of action, researchers can better identify the factors for which measures are needed in order to study a phenomenon adequately. The TBEST measures one variable in a complex equation.

Developing a beliefs measure should be a collaborative effort. As we described in the chapter, group item writing brought diverse viewpoints to bear on the task, often revealing unexamined assumptions. Exposing these assumptions in the development phase forced us to confront ambiguous wording and made it more likely that survey items would be interpreted as intended by respondents. The group process also served as an ongoing check on face validity of the items.

Development should include cognitive interviews with the target audience. Even a thorough collaborative development effort cannot anticipate all of the ways in which teachers will interpret questionnaire items. Cognitive interviews, in which teachers think aloud as they respond to items, are critical for eliminating sources of unreliability and invalidity. Interviews on the TBEST items, for instance, revealed that teachers interpreted the word "phenomena" quite differently than the developers intended.

Development should incorporate multiple rounds of piloting with the target audience. It is likely that the first round of piloting will identify items that should be deleted, and new items may be needed to fill in gaps or to shore up factors that have an insufficient number of items. In either case, further piloting is warranted. In developing the TBEST, we conducted four pilot studies, both to test new versions of the survey and to explore the survey's robustness across audiences and administration modes.

The field has a wealth of analytical tools for exploring data that result from pilot studies. Using these tools is both an art and a science; it requires individuals with a firm command of the tools who can think in terms of tradeoffs. The analysis process used in developing the TBEST included multiple decision points, some of which we anticipated and were able to address in advance. Others depended on the interpretation of initial models to provide guidance for the survey design as well as subsequent analyses.

Finally, developing a beliefs measure takes a considerable amount of time. The TBEST development spanned several years. *Developers will make decisions throughout, and it is important that these be documented so that the history can be accurately constructed.* The decisions and accompanying rationales are important for describing the instrument to potential users, and they contain important lessons for other developers.

REFERENCES

Ajzen, I. (1985). From intentions to actions: A theory of planned behavior. In J. Kuhl & J. Beckman (Eds.), *Action-control: From cognition to behavior* (pp. 11–39). Heidelberg, Germany: Springer.

Ajzen, I. (2012). The theory of planned behavior. In P. A. M. Lange, A. W. Kruglanski, & E. T. Higgins (Eds.), *Handbook of theories of social psychology: Vol. 1.* (pp. 438–459). London, UK: Sage.

Ajzen, I., & Gilbert Cote, N. (2008). Attitudes and prediction of behavior. In W. D. Crano & R. Prislin (Eds.), *Attitudes and attitude change* (pp. 289–311). New York, NY: Psychology Press.

American Association for the Advancement of Science. (1993). *Benchmarks for science literacy : Project 2061.* New York, NY: Oxford: Oxford University Press.

Bandura, A. (1997). *Self-efficacy: The exercise of control.* New York, NY: W.H. Freeman.

Banilower, E., Cohen, K., Pasley, J., & Weiss, I. (2008). *Effective science instruction: What does research tell us.* Portsmouth. NH: Center for Instruction.

Browne, M. W., & Cudeck, R. (1993). Alternative ways of assessing model fit. *Sage Focus Editions, 154,* 136–136.

Cobern, W. W., & Loving, C. C. (2002). Investigation of preservice elementary teachers' thinking about science. *Journal of Research in Science Teaching, 39*(10), 1016–1031.

Desimone, L. M., & Le Floch, K. C. (2004). Are we asking the right questions? Using cognitive interviews to improve surveys in education research. *Educational Evaluation and Analysis, 26*(1), 1–22. Doi:10.3102/01623737026001001

Lederman, N. G., Abd□El□Khalick, F., Bell, R. L., & Schwartz, R. S. (2002). Views of nature of science questionnaire: Toward valid and meaningful assessment of learners' conceptions of nature of science. *Journal of Research in Science Teaching, 39*(6), 497–521. Doi:10.1002/tea.10034

Lumpe, A. T., Haney, J. J., & Czerniak, C. M. (2000). Assessing teachers' beliefs about their science teaching context. *Journal of Research in Science Teaching, 37*(3), 275–292.

Magnusson, S., Krajcik, J., & Borko, H. (1999). Nature, sources and development of pedagogical content knowledge for science teaching. In J. Gess-Newsome & N. G. Lederman (Eds.), *Examining pedagogical content knowledge* (pp. 95–132). Norwell, MA: Kluwer Academic Publishers.

National Research Council. (1996). *National science education standards: Observe, interact, change, learn.* Washington, DC: National Academy Press.

National Research Council. (2000). *How people learn: Brain, mind, experience, and school (Expanded Edition).* J. D. Bransford, A. L. Brown, & R. R. Cocking (Eds.), Washington, DC: National Academy Press.

National Research Council. (2005). *How students learn: Science in the classroom.* M. S. Donovan & J. D. Bransford (Eds.), Washington, DC: National Academy Press.

Posner, G. J., Strike, K. A., Hewson, P. W., & Gertzog, W. A. (1982). Accommodation of a scientific conception: Toward a theory of conceptual change. *Science education, 66*(2), 211–227.

Riggs, I. M., & Enochs, L. G. (1990). Toward the development of an elementary teacher's science teaching efficacy belief instrument. *Science Education, 74*(6), 625–637.

Sampson, V., & Benton, A. (2006). *Development and validation of the beliefs about reformed science teaching and learning (BARSTL) questionnaire.* Annual Conference of the Association of Science Teacher Education (ASTE). Portland, Oregon. Retrieved February (Vol. 16, p. 2008).

Schumacker, R. E., & Lomax, R. G. (1996). *A beginner's guide to structural equation modeling.* Mahwah, NJ: L. Erlbaum Associates.

Smolleck, L. D., Zembal-Saul, C., & Yoder, E. P. (2006). The development and validation of an instrument to measure preservice teachers' self-efficacy in regard to the teaching of science as inquiry. *Journal of Science Teacher Education, 17*(2), 137–163.

Tabachnick, B. G., & Fidell, L. S. (2007). *Using multivariate statistics.* Boston: Pearson/Allyn & Bacon.

NOTES

[1] Magnusson, Krajcik, and Borko (1999) provide a thorough treatment of the distinction between subject matter knowledge and pedagogical content knowledge for science teaching. In their framework, subject matter knowledge contributes to but is distinct from pedagogical content knowledge.

[2] Other chapters in this book describe competing and complementary theories regarding the relationship between beliefs and actions.

[3] Actual behavioral control is the extent to which the individual has what is needed in order to engage in a behavior, in contrast to the individual's perceptions of the presence of these factors. Perceived behavioral control may or may not align closely with actual behavioral control. If the alignment is close, perceived control serves as a proxy for actual control in predicting behavior, as represented by the dashed line in Figure 2.

[4] There are five versions of the V-NOS for various audiences, consisting of between six and ten questions.

[5] In order to explore relationships between beliefs and other factors, the instruments must be sensitive to variation in the constructs of interest.

[6] Using Direct Oblimin in SPSS version 19.

[7] These fit indices are typically referred to in abbreviated form. The formal names of the fit indices are: CFI = Comparative Fit Index; TLI = Tucker-Lewis Index; RMSEA = Root Mean Square Error of Approximation. For more information about each fit index, see Tabachnick & Fidell, (2007).

AFFILIATION

P. Sean Smith
Senior Researcher & Partner
Horizon Research, INC

Adrienne A. Smith
Research Associate
Horizon Research, INC

Eric R. Banilower
Senior Researcher & Partner
Horizon Research, INC

APPENDIX

TEACHER BELIEFS ABOUT EFFECTIVE SCIENCE TEACHING (TBEST)
QUESTIONNAIRE

Questionnaire Instructions:

This questionnaire asks you to respond to 21 statements regarding your beliefs about effective science instruction; that is, what does science instruction that helps students learn science concepts well look like?

Teachers have to make many trade-offs when they are responsible for teaching many standards in one year. Teachers may not be able to emphasize the instructional strategies they believe are effective and still cover the entire curriculum. When you respond to the statements below, please try to put those trade-offs aside. Imagine that you are not constrained by state/district standards, or available time/resources, or feasibility issues. What does effective science instruction look like, without all the constraints that limit what you can do in the classroom.

When responding to the statements, please try to think about students in general, not one student or a particular group of students.

Finally, this questionnaire makes frequent use of two terms that teachers may interpret differently depending on the context. For the purpose of this questionnaire, please use the following definitions of "data" and "evidence."

Data—information that has not yet been analyzed or processed; typically gathered through observation or measurement.

Evidence—analyzed or processed data that are used to support a scientific claim or conclusion.

These definitions are repeated on each page of the questionnaire.

TBEST QUESTIONNAIRE

Practical constraints aside, do you agree that doing what is described in each statement would help most students learn science?

Circle one in each row.

	Strongly Disagree	Moderately Disagree	Slightly Disagree	Slightly Agree	Moderately agree	Strongly Agree
1 At the beginning of instruction on a science concept, students should be provided with definitions for new scientific vocabulary that will be used.	1	2	3	4	5	6
2 Hands-on activities and/or laboratory activities should be used primarily to reinforce a science concept that the students have already learned.	1	2	3	4	5	6
3 Students should rely on evidence from classroom activities, labs, or observations to form conclusions about the science concept they are studying.	1	2	3	4	5	6
4 Teachers should have students do hands-on activities, even if the data they collect are not closely related to the concept they are studying.	1	2	3	4	5	6
5 Teachers should explain a concept to students before having them consider evidence that relates to the concept.	1	2	3	4	5	6
6 Teachers should provide students with opportunities to connect the science they learn in the classroom to what they experience outside of the classroom.	1	2	3	4	5	6
7 Teachers should ask students to support their conclusions about a science concept with evidence.	1	2	3	4	5	6
8 Students should do hands-on or laboratory activities, even if they do not have opportunities to reflect on what they learned by doing the activities.	1	2	3	4	5	6

	1	2	3	4	5	6
9 At the beginning of instruction on a science concept, students should have the opportunity to consider what they already know about the concept.	1	2	3	4	5	6
10 Students should do hands-on activities after they have learned the related science concepts.	1	2	3	4	5	6
11 Teachers should provide students with opportunities to apply the concepts they have learned in new or different contexts.	1	2	3	4	5	6
12 Students should use evidence to evaluate claims about a science concept made by other students.	1	2	3	4	5	6
13 Teachers should have students do interesting hands-on activities, even if the activities do not relate closely to the concept being studied.	1	2	3	4	5	6
14 At the beginning of lessons, teachers should 'hook' students with stories, video clips, demonstrations or other concrete events/activities in order to focus student attention.	1	2	3	4	5	6
15 Students' ideas about a science concept should be deliberately brought to the surface prior to a lesson or unit so that students are aware of their own thinking.	1	2	3	4	5	6
16 Teachers should provide students with the outcome of an activity in advance so students know they are on the right track as they do the activity.	1	2	3	4	5	6
17 Students should have opportunities to connect the concept they are studying to other concepts.	1	2	3	4	5	6

(*Continued*)

	Circle one in each row.					
	Strongly Disagree	*Moderately Disagree*	*Slightly Disagree*	*Slightly Agree*	*Moderately agree*	*Strongly Agree*
18 Students should consider evidence that relates to the science concept they are studying.	1	2	3	4	5	6
19 When students do a hands-on activity and the data don't come out right, teachers should tell students what they should have found.	1	2	3	4	5	6
20 Students should know what the results of an experiment are supposed to be before they carry it out.	1	2	3	4	5	6
21 Students should consider evidence for the concept they are studying, even if they do not do a hands-on or laboratory activity related to the concept.	1	2	3	4	5	6

GAIL SHROYER, IRIS RIGGS & LARRY ENOCHS

MEASUREMENT OF SCIENCE TEACHERS' EFFICACY BELIEFS

The Role of the Science Teaching Efficacy Belief Instrument

INTRODUCTION

Since its development in 1990, the Science Teaching Efficacy Belief Instrument (STEBI, versions A and B) has been used to investigate pre-service and in-service elementary teachers' beliefs about teaching and learning science from multiple international perspectives. This chapter will consider the ways that the STEBI has been used by educational researchers, the contributions STEBI-based research has made to our understanding of science teaching self-efficacy, the potential benefits to the field of science education reaped by investigation of teacher beliefs, and recommendations for future STEBI-supported research.

The STEBI is a self-report instrument in which teachers rate their agreement with items regarding their ability to teach science (self-efficacy) and also their belief in students' ability to learn science (outcome expectancy) through response to a 5-point scale that ranges from strongly agree to strongly disagree. Perhaps the ease of implementation of the STEBI contributes to its frequent use in studies that range from dissertations to evaluations of funded projects and pre-service programs across different cultural contexts. Teachers or teacher candidates can respond to the survey in approximately 15 minutes, and analysis is quick and easy--thus resulting in a quantitative research approach that is inexpensive and efficient.

Still, there are enough STEBI-based studies that reach beyond simple documentation of self-report results to suggest that the constructs that the instrument measures have prompted the science education community to reflect upon the role that teacher beliefs play within the professional development process. The STEBI's roots lie within a study of general teacher self-efficacy done by Gibson and Dembo in 1984. Gibson and Dembo developed a 30-item scale that resulted in two types of teacher beliefs that they related to constructs identified by Albert Bandura within his social cognitive theory--the constructs of self-efficacy and outcome expectancy (Bandura, 1977). When applied to teachers, Gibson and Dembo defined self-efficacy as teachers' beliefs in their own ability to teach, while outcome expectancy was seen as teachers' beliefs in students' ability to learn despite other environmental factors such as school context or student background (Gibson & Dembo, 1984).

R. Evans et al., (Eds.), The Role of Science Teachers' Beliefs in International Classrooms, 103–118.
© *2014 Sense Publishers. All rights reserved.*

Ultimately, the STEBI instrument was modeled after the Gibson & Dembo *Teacher Efficacy Scale* (TES). In the case of elementary teachers, who have responsibility for teaching all content areas, the constructs identified in the TES were seen as having great potential for study of elementary teachers' responses to teacher preparation or professional development in science. For example, what type of professional development has the most impact on how much time elementary teachers dedicate to teaching science? Are those teachers with higher levels of self-efficacy more likely to spend more time on science? Can professional development positively impact teacher self-efficacy beliefs, and therefore also impact positive instructional change? Do some elementary teachers avoid teaching much science even if they have high self-efficacy for teaching it, due to their belief that students aren't capable of learning science, perhaps because of limited background knowledge, etc? Can and should teacher development attempt to address teacher beliefs about student potential for learning in order to maximize benefit gleaned from professional development?

If self-efficacy and outcome expectancy beliefs are truly related to teacher behaviors in the classroom as was suggested by Gibson and Dembo and others, then it seems that a similar measure specific to elementary science teachers would contribute to the field by helping researchers more easily evaluate their attempts to prepare and develop elementary teachers' ability to teach science to students.

SCIENCE TEACHING EFFICACY BELIEF INSTRUMENT

STEBI Development

The STEBI emerged from work being done beginning in 1984 at Kansas State University involving secondary science teachers being prepared to lead district-wide science improvement efforts in rural schools (NSF TEI 84-70338). As the project staff studied factors involved in school change, several variables emerged. These included teacher beliefs as studied within the Rand study (Berman, McLaughlin, Bass, Pauly, & Zellman, 1977). This early interest in teacher beliefs was further supported by Gibson and Dembo (1984) and within work by Ashton and Webb (1986) in the area of self-efficacy of elementary teachers.

The first instrument to measure teacher efficacy originated in the RAND study that evaluated 100 Title III projects associated with the 1965 Elementary and Secondary Education Act (Berman, McLaughlin, Bass, Pauly, & Zellman, 1977). These researchers used two efficacy items and investigated teachers' beliefs concerning aspects the teachers felt they could control (or at least strongly influence) in their school and their students' achievement and motivation. The results pointed to efficacy as the most important factor when it came to teachers promoting their students' learning and motivation. These results were confirmed in studies on implementation of new school or district programs (Ashton & Webb, 1986).

The two items from the RAND study proved so powerful in predicting student performance, teacher change, and continued use of methods and materials from federally funded projects that many different multiple item instruments were developed to capture teacher efficacy. The two items were based on Rotter's (1966) locus of control theory (Armor, Conroy-Oseguera, Cox, King, McDonnell, Pascal, Pauly, & Zellman, 1976):

Item 1: "When it comes right down to it, a teacher really can't do much because most of a student's motivation and performance depends on his or her home environment."

Item 2: "If I really try hard, I can get through to even the most difficult or unmotivated students."

The items were intended to assess whether a teacher believed that student learning and motivation were under the teacher's control. These items and this locus of control orientation guided most teacher efficacy research during the late 70s and early 80s.

One of the most widely used instruments was the Gibson and Dembo (1984) instrument, which was based on the RAND study items but utilized the framework of self-efficacy from social cognitive theory (Bandura, 1986). Utilizing this instrument, research studies with elementary, middle, and high school teachers demonstrated that teachers' self-efficacy beliefs impacted student outcomes and teacher behaviors. In a review of the literature on teacher self-efficacy, Tschannen-Moran, Hoy, and Hoy (1998) indicated teachers' self-efficacy beliefs were related to student outcomes such as achievement, motivation, and the students' sense of efficacy. They also indicated teaching self-efficacy was related to teacher classroom behaviors, the goals set, persistence with students, and enthusiasm for and commitment to teaching. Teachers with high teaching self-efficacy performed better and their students benefited.

Gibson and Dembo (1984) extended the research and found that self-efficacy was comprised of two separate and uncorrelated factors: factors they called personal teaching efficacy and teaching outcome expectancy. However, efficacy is a situation and context specific construct (Bandura, 1986). Teachers may feel quite efficacious concerning some classes and feel quite the opposite in other content areas. A general efficacy instrument, like Gibson and Dembo's, proved to be ineffectual at capturing these content specific situations. Bandura (1986) cautioned that, because judgments of self-efficacy are task- and domain-specific, "ill-defined global measures of perceived self-efficacy or defective assessments of performance would yield discordances" (p. 397).

The Science Teaching Efficacy Beliefs Instrument (STEBI) was modeled after the Gibson & Dembo Teacher Efficacy Scale (TES). Thus, its items were modified to include the elementary science classroom setting. The instrument was entitled the Science Teaching Efficacy Belief Instrument (STEBI) and included two subscales. Initially, a large item pool was created. An educational measurement expert edited these items for clarity. The resultant 50 items were submitted to a panel of experts in the field for construct validity. The panel was asked to use the operational definitions for the two constructs to assess items' meaning and clarity and to classify them into two subscales. The following is an example of an outcome expectancy item (STOE): "When a student does better than usual in science, it is often because the teacher exerted a little extra effort." An example of a personal self-efficacy item (PSTE) is: "I am continually finding better ways to teach science."

The resultant instrument was administered in a pilot study involving 71 practicing elementary teachers enrolled in graduate courses. Reliability analysis of the Personal Science Teaching Efficacy Scale produced an alpha of 0.92. The Science Teaching Outcome Expectancy scale reported an alpha of 0.73. Factor analysis for the revised scale showed all items correlating highly with their own scale (Riggs & Enochs,1990).

The refined Science Teaching Efficacy Belief Scale was administered to a new and larger sample of practicing elementary teachers (N = 331), both rural and urban. A one-tailed t-test was used to ensure that no significant differences existed between rural and urban samples for both scales. Instrument reliability was again estimated through the internal consistency procedure described previously. Items that did not have a high positive discrimination index were rejected. Initial factor analysis was used to determine the number of significant factors. Two factors produced eigenvalues >1.00. A second factor analysis, limited to the final number of factors (2), was used to ensure appropriate scale loadings. Items that cross-loaded or loaded into the wrong factor were eliminated.

A majority of the respondents were white and female. All elementary grade levels were represented as well as varied levels of teacher experience. Rural and urban teachers were also included in the sample with no significant difference between the two sub-groups identified by t-tests. Additional t-tests were run on the scale scores of all other demographic characteristics. Only gender exhibited a significant difference with higher scores found for males on the Personal Science Teaching Efficacy Belief scale at the 0.05 level. Item analysis was again conducted on both scales. For the Personal Science Teaching Efficacy Belief scale, an alpha of 0.91 was achieved. All items corrected item-total correlations of 0.53 and above except for two. These were deleted, increasing the balance of item phrasing in this scale and raising alpha to 0.92. Factor analysis supported the contention that the scales were distinct and measurable constructs. As predicted by social learning theory, a modest correlation was found between the two subscales.

The STEBI has since been adapted for use with chemistry teachers (STEBI-CHEM: Rubeck & Enochs, 1991), mathematics teachers (MTEBI: Enochs, Smith, & Huinker), outdoor educators (Holden, Grouix, Bloom, & Weinburgh, 2011), teachers completing HIV/AIDS intervention modules (Webb & Gripper, 2010), and to measure prospective elementary teachers' self-efficacy beliefs about equitable science teaching and learning (SEBEST: Ritter, Boone, & Rubba, 2001). The STEBI also has been used, translated, and adapted for use in Australia (Ginns & Watters, 1999), Denmark (Andersen, Dragsted, Evans, & Sorensen, 2003 as cited in Mihladiz, Duran, Isik, & Ozdemir, 2010), South Africa (Webb & Gripper, 2010), Turkey (Mihladiz et al, 2010), and Singapore (Wee-Loon, 2011).

Uses of STEBI in Teacher Education

Since the development of the STEBI A for in-service teachers (Riggs & Enochs, 1990) and the STEBI B for pre-service teachers (Enochs & Riggs, 1990), these instruments have been used internationally to help educators understand and enhance teacher education. Teacher educators have compared STEBI B scores across different groups of students exposed to innovative science courses, methods courses, field experiences, and student teaching approaches to measure the impact of these reformed teacher education practices on personal science teaching self-efficacy (PSTE) and science teaching outcome expectancy (STOE) beliefs of future teachers. Researchers have examined the development of self-efficacy beliefs of both pre-service and in-service teachers in relation to science knowledge, the number of science courses taken in high school and college, and other antecedent life experiences. These results are often compared to Bandura's theory of self-efficacy and his four sources of efficacy information. Researchers also have explored the ramifications of self-efficacy beliefs by comparing both pre-service and in-service STEBI scores with preferences for teaching, teaching approaches, self-rated teaching effectiveness, and observed teaching effectiveness. The majority of studies have combined quantitative analysis of STEBI scores in relation to scores from other instruments, interviews, observations, and rich descriptions of both pre-service and in-service teacher education program elements.

Many authors have documented increases in STEBI B scores after pre-service teachers' exposure to innovations in teacher education (Cantrell, Young, & Moore, 2003; Liang & Richardson, 2009; Mulholland, Dorman, & Odgers, 2004; Palmer, 2006; Perkins, 2007; Richardson & Liang, 2008; Settlage, Southerland, Smith, & Ceglie, 2008; Tosum, 2000; Young & Kellogg, 1993). On the other hand, methods courses are not always associated with increases in self-efficacy beliefs (Bursal, 2008). Ginns and Watters (1999), using a multiple-case study approach involving beginning teachers in Queensland, documented decreases in STEBI scores from pre-service experiences to first year teaching experiences with increases during the second year of teaching. Shroyer (1997) and Wilson (1996) reported mixed results from a longitudinal study of reform in teacher education, depending upon the nature of the innovation being examined.

Teacher educators have documented increases in science teaching self-efficacy scores when their students participated in revised methods courses aligned with the science standards (Christol & Adams, 2006), constructivist oriented methods courses (Bleicher & Lindgreen, 2005), cooperative learning teaching methods experiences (Scharmann & Hampton, 1995), cooperative learning focused field experiences (Cannon & Scharmann, 1995), and inquiry-based experiences during methods courses (Liang & Richardson, 2009; Palmer, 2006; Richardson & Liang, 2008). These studies all appeared to revolve around well-designed and sequenced methods experiences.

Personal science teaching efficacy (PSTE) scores also appeared to increase as future teachers were given opportunities to teach science. Cantrell, Young, and

Moore (2003) documented higher self-efficacy scores among pre-service teachers who taught science to children more than three hours a week. An investigation of 492 pre-service teachers in Turkey indicated those in their final year had significantly higher self-efficacy beliefs possibly due to the teaching experience course taken by seniors (Aydin & Boz, 2010). Shroyer (1997) reported a 10-year longitudinal study of reform at one institution indicated a consistently positive relationship between group STEBI B scores and increased field experiences. This research, supported by interviews and teaching observations, also indicated that success during field experiences may influence PSTE scores more than number of teaching episodes. When teaching experiences were not successful, individual PSTE scores declined or remained low whether the student had the opportunity to teach five lessons or only one lesson (Shroyer, 1997).

As a point of comparison, outcome expectancy beliefs (STOE) of pre-service teachers were more difficult to interpret and not as easily changed. While some researchers reported significant changes in both PSTE and STOE, (Bleicher & Lindgreen, 2005; Shroyer, 1997) more have reported significant changes in PSTE scores without changes in STOE scores (Cantrell, Young, & Moore, 2003; Liang & Richardson, 2009; Schoon & Boone, 1998, Tosun, 2000). Ginns, Watters, Tulip, and Lucas (1995) reported significant changes in STOE without changes in PSTE. According to these authors, "We do not see these mixed result as inconsistent. We feel they show that different interventions in pre-service teacher preparation courses can result in changes to either self-efficacy or outcome beliefs, and sometimes to both" (p. 218). Others argued, "since STOE interpretations have been problematic, further studies are needed to determine how STOE should be conceptualized, operationalized, and measured," (Cone, 2009, p. 381).

A good deal of research has been conducted to determine correlational and antecedent experiences related to science teaching self-efficacy of pre-service as well as in-service teachers using both the STEBI A and B. Researchers have documented that personal science teaching self-efficacy (PSTE) was positively related to science content knowledge (Lloyd, Smith, Fay, Khang, Wah, & Sai, 1998; Stevens & Wenner, 1996), conceptual understanding of science (Bleicher, 2006; Bleicher & Lindgren, 2005; Perkins, 2007), fewer alternative conceptions in science (Schoon & Boone, 1998), and science laboratory competencies (Mihladiz, Duran, Isik, & Ozdemir, 2010).

The relationship between science teaching self-efficacy beliefs and science course taking patterns was less conclusive. Science teaching self-efficacy beliefs have been positively related to the number of high school and college classes taken and involvement in high school extracurricular science activities (Bleicher, 2004; Cantrell, Young, & Moore, 2003; Ginns et al, 1995), coursework with laboratory experiences (Rubeck & Enochs, 1991) and educational degree level (Ramey-Gassert, Shroyer, & Staver, 1996), but others have documented no statistical correlation between number of science courses taken and science teaching efficacy (Bleicher & Lindgren, 2005; Stevens & Wenner, 1996) or even a negative correlation between number of high school and college courses taken and personal science teaching self-efficacy (PSTE) (Enochs, Scharmann, & Riggs, 1995). Perkins (2007) noted differences between studies might be based on differences between reformed and traditional college science courses.

Since the relationship between understanding science and science teaching self-efficacy has been more clearly demonstrated, one interpretation is that not all high school and college science courses lead to increased understanding of science, particularly for elementary teachers. Over a period of five years in the 1990s, one institution initiated and or revised several content courses for elementary teachers (Shroyer, 1997; Wilson, 1996). These courses were in biology, earth science, chemistry, and physics. They all included hands-on experiences and lectures with differing levels of inquiry. Regardless of the design of the course, PSTE scores were related to students' perceptions of their own ability to understand the concepts they were being taught. If students struggled with the concepts they were learning, PSTE scores indicated they were more likely to believe they would not be successful teaching these concepts to children. Learning challenging concepts like physics, even when taught through well-designed courses that included weekly labs designed to help students understand the concepts, did not lead to increases in PSTE scores for students who struggled with the concepts themselves.

International comparisons of teacher efficacy beliefs have demonstrated cross-cultural differences that suggest these beliefs may be influenced by culture as well as differences between coursework, field experiences, and characteristics of the teachers being studied due to teacher education entrance criteria (Cakiroglu, Cakiroglu, & Boone, 2005). A modified version of the STEBI was used in South Africa, for example, to assess changes in beliefs of 128 teachers participating in an HIV/AIDS module (Webb & Gripper, 2010). Although significant increases were revealed in pre- and post-test STEBI scores related to understanding the topics of study, cultural barriers inhibited teachers' confidence in their ability to implement an HIV/AIDS educational program.

Andersen, Dragsted, Evans, and Sorensen (2003) as cited in Mihladiz, Duran, Isik, and Ozdemir (2010) translated as well as adapted the STEBI for use in Denmark based on differing cultural values and school norms. The new instrument (the STEBI – DK) demonstrated similar reliability compared to the original STEBI. The STEBI-DK was tested with pre-service students in Denmark and the USA. Cultural, attitudinal, and preparatory differences between Danish and American pre-service students were evident. The pre-service students from North Carolina showed significantly higher personal science teaching (PSTE) as well as outcome expectancy (STOE) scores compared to the pre-service teachers from Copenhagen (Mihladiz, Duran, Isik, & Ozdemir, 2010).

Pre-service teachers from Turkey demonstrated average (Azar, 2010; Mihladiz, Duran, Isik, & Ozdemir, 2010) to high (Cakiroglu et al, 2005; Yilmaz & Cavas, 2008) self-efficacy beliefs while comparisons to pre-service students in the USA revealed similarities as well as differences in beliefs (Cakiroglu et al, 2005).

Pre-service teachers from the Midwest had significantly stronger personal science teaching efficacy beliefs (PSTE) compared to Turkish pre-service teachers, but there was no significant difference between their beliefs that their teaching can influence student learning (STOE). Preparatory differences were again noted. Pre-service teachers from Turkey were more likely to believe they would welcome and be able to answer students' science questions while those from the USA were more likely to believe they would be able to help students with difficulties in understanding science.

Clearly, there were numerous cultural, educational, and other life experiences that influenced self-efficacy beliefs. Many researchers have used interviews with and written reflections from pre-service as well as in-service teachers along with scores from the STEBI and other instruments to enhance their understanding of the factors that teachers perceive have influenced their science self-efficacy. Such studies frequently compared perceptions from teachers to Bandura's four sources of efficacy information: mastery experiences, social persuasion, vicarious experiences, and physical and emotional states.

As predicted by social learning theory, mastery experiences, primarily represented as successful experiences learning and teaching science, were the most well documented factors teachers believed have impacted their self-efficacy (Aydin & Boz, 2010; Ginns & Waters, 1999; Liang & Richardson, 2009; Perkins, 2007; Ramey-Gassert et al, 1996; Shroyer, 1997). Social persuasion also had been well documented as having a powerful affect on self-efficacy through social verbal support (Perkins, 2007), positive feedback and enthusiastic responses from the students being taught (Ginns & Watters, 1999; Wee-Loon, 2011), a perceived supportive learning environment, (Liang & Richardson, 2009), district level community support (Rubeck & Enochs, 1991), and supportive supervisors and colleagues who inspire and assist (Wee-Loon, 2011).

Liang and Richardson (2009) indicated the importance of vicarious experiences involving peer problem solving and cooperative learning while Aydin and Boz (2010) described the vicarious experiences associated with former teachers and classroom observations. The impact of physical and emotional states on self-efficacy beliefs was related to stress reduction, a sense of meaningfulness, relevance, enjoyment (Liang & Richardson, 2007), and a life-long passion for science among Singapore teachers (Wee-Loon, 2011). Ramey-Gassert et al (1996) found positive correlations between STEBI scores and attitudes toward science while Bleicher (2004) reported positive correlations between self-efficacy beliefs and positive school science experiences.

The STEBI also has been used to explore the relationship between science teaching self-efficacy beliefs and pre-service and in-service teaching practices. Enochs, Scharmann, and Riggs (1995) found significantly positive correlations between science teaching self-efficacy beliefs of pre-service teachers, choosing to use activity-based science instruction, and perceived effectiveness in teaching science. These authors also reported a correlation between self-efficacy and pupil control orientation and hypothesized those with high efficacy beliefs would be less authoritarian teachers. Czerniak & Lumpe (1996) noted significantly positive correlations between in-service teachers' self-efficacy and their beliefs that reform recommendations are necessary to be an effective teacher. Ramey-Gassert et al (1996) reported a positive correlation between personal science teaching efficacy (PSTE) and choosing to teach science as well as self-rated effectiveness in teaching science among practicing teachers.

Some studies have indicated a positive relationship between teacher beliefs, as measured by the STEBI, and observed teaching behaviors (Riggs, Enochs, & Posnanski, 1998). Haney, Lumpe, Czerniak, and Egen (2002) cautiously noted a positive relationship between self-efficacy beliefs and the actions of practicing teachers. Generally teachers with higher efficacy beliefs scored higher on observed effective classroom practices; although there was one exception, out of six teachers studied, of a teacher with high self-efficacy beliefs and "substantial problems observed in her implementation, content knowledge, and classroom environment" (p. 181).

Other researchers also have documented high levels of self-efficacy that were incongruous with ability to teach. Lardy and Mason (2011) conducted a large national study of 85 in-service teachers from diverse backgrounds trained in "reformed", inquired-oriented teacher education programs. These participants were observed during their first few years of teaching using an observational protocol designed to assess reformed teaching. There was no correlation between the participants' observed teaching and their self-efficacy scores. The authors concluded, "Therefore, a blanket assumption can not be made that increasing the self-efficacy beliefs of pre-service and in-service elementary teachers will automatically improve their ability to effectively teach science to their students. The relationship between science teaching self-efficacy beliefs and science teaching behaviors is much more complex than we might assume" (p. 21). Unfortunately, teachers with high STEBI scores and low observed reformed teaching behaviors believed they were teaching effectively. Settlage et al (2008) spoke to the blinding effect high efficacy can have on teaching performance. These authors pointed to the value Dewey placed on "uncertainty" and, citing Wheatley (2002), viewed self-doubt as a force to foster professional growth.

There was evidence that the STEBI also had been used to evaluate professional experiences offered in informal settings such as science centers or museums. For example, in Australia, McKinnon (2010) required both pre-service and in-service teachers to complete the STEBI immediately after, four months following, and eleven months after completing a series of four workshops offered by a science center. Self-efficacy results were analyzed and related to school context, reform efforts, and the role that informal education might play in promoting positive change in science teaching.

American researchers used the STEBI to investigate the impact of professional development in outdoor education on science teachers' self-efficacy beliefs (Holden, Grouix, Bloom, & Weinburgh, 2011). The STEBI was modified to include "outdoor" as a modifier to each reference of science teaching or learning. Findings prompted the authors to consider issues related to teachers' ability to transfer learning from the outdoor setting back to their classrooms. They also discussed potential weakness in the STEBI's ability to measure teacher beliefs given the context of today's schools. Citing, Wheatley (2005), Holden et al (2011) stated:

> The STEBI has served science education very well for many years. However, in today's classroom environments and considering the evolved view of what constitutes quality science teaching, we must ask ourselves if the STEBI instrument can still adequately measure the perceived self-efficacy of today's teachers. Perhaps the STEBI (and other efficacy instruments of the same era) does not accurately measure either the reality of accountability at-all-costs of No Child Left Behind or the democratic, inquiry-oriented classroom to which many science teachers and science teacher educations currently ascribe. (Wheatley, 2005)

We agreed that the STEBI's ability to continue to serve today's researchers deserved consideration and do so in the following section.

RECOMMENDATIONS FOR FUTURE STEBI RESEARCH

Clearly, the STEBI, and instruments like it, can contribute to the research community by providing an easily implemented and analyzed assessment of teacher beliefs. Measures can serve as one, simplistic indicator of teacher beliefs prior to, during, and after methods courses, field experiences, or professional development, thus indicating some level of evaluation of interventions. Results can be compared to other variables such as teacher experience and school or student contexts, thus leading to further understanding of the interplay between the nature of teachers' backgrounds, the characteristics of their school contexts, and the challenges their students face along with teachers' efforts to implement changes advocated through teacher education and professional development.

Still, one must question why the STEBI, in its original form, continues to be used to such an extent, more than two decades after its publication. Yes, many researchers have modified the STEBI to a particular context or culture, but typically these modifications have been minor (as stated previously, adding "outdoor" as a modifier of "science teaching"). Even when translated or modified, STEBI items tend to measure teachers' beliefs about science teaching and learning in general terms with little mention of what science teaching entails except for a few items that reference such practices as welcoming student questions or monitoring science experiments. Perhaps some might say that the generality of the STEBI items has allowed it to be useful over time because teachers can apply their own definitions of science teaching and learning as they respond to the items. This level of generality has allowed it to be used internationally in varied cultural settings.

Certainly, should researchers continue using the STEBI, they need to continue to collect additional in-depth information from, about, and with teachers, in order to more adequately interpret STEBI results. Henson (2002) suggests researchers consider context more directly through observations of classroom practices and teacher "think alouds". Wheatley (2005) advocates refocusing self-efficacy research on teachers' interpretations and involving teachers in the research. As far back as 1992, Pajares called for additional qualitative methodologies such as case studies, oral histories, and the use of metaphor, biography, and narrative to more deeply understand the beliefs of teachers. The strongest research cannot solely rely upon a single, self-reported measure of teacher beliefs.

It also seems appropriate to consider whether STEBI items might be reconsidered given advances that have been made in defining expectations for good science teaching and learning (NRC, 2000). The Teaching Science as Inquiry Instrument (TSI), developed by Smolleck, Zembal-Saul, and Yoder (2006), is one attempt at creating a more specific self-efficacy measure. The TSI's authors based its items on the The Five Essential Features of Classroom Inquiry as advocated by the National Research Council (2000). The measure moves beyond the STEBI's general measure to have teachers reflect on their ability to promote their students' ability to engage in the five characteristics of science inquiry--asking scientifically oriented questions, prioritizing evidence in responding to questions, developing explanations based upon evidence, connecting explanations to scientific knowledge, and articulating and justifying explanations (NRC 2006; Smolleck et al, 2006).

Bandura proposes that self-efficacy beliefs are specific, yet, specific measures, depending on their degree of specificity, might include their own set of challenges. As Woolfolk Hoy (2000) advises,

> In order to be useful and generalizable, measures of teacher efficacy need to tap teachers' assessments of their competence across the wide range of activities and tasks they are asked to perform. And yet there is a danger of developing measures that are so specific they loose their predictive power for anything beyond the specific skills and contexts being measured" (p. 9).

Specific measures often require a level of professional language or jargon that could inhibit respondents' ability to fully comprehend the items. Thus, a negative response might be due not to lower self-efficacy and outcome expectancy beliefs, but instead could be the result of misunderstanding of the specific dimensions of the instrument.

Developers of specific measures are advised to be mindful of item readability. For example, if pre-service teachers respond to a measure prior to taking a course in which they develop understanding of the instrument's language and concepts, then post-assessment results most likely are influenced by increased comprehension of items in addition to actual changes in beliefs. Too much specificity most certainly would limit an instrument's ability to be used cross-culturally; whereas the STEBI's level of specificity appears to have allowed it to contribute to understanding of teachers' beliefs about science teaching and learning across the globe.

Whether researchers adapt the STEBI to be more specific or simply continue to use it in its present form, whenever teacher beliefs are measured through self-report surveys such as the STEBI, we find the recommendations of Boone, Townsend, and Staver (2011) serve as a valuable guide. The authors provide comprehensive directions on using Rasch analyses to guide development of instruments like the STEBI, to make the most of rating scale-based data, and to connect theory to instrumentation. Their comprehensive work utilized the STEBI as an example of how researchers might adjust an already existing instrument through careful use of Rasch.

Other instrumentation issues worthy of further research include the need for stronger theoretical models of self-efficacy to clarify the differences between outcome expectancy, for example, and locus of control (Henson, 2002; Tschannen-Moran et al., 1998). In response to critical theorists, Labone (2004) encourages "broadening the construct of teacher efficacy to explore dimensions of efficacy that facilitate educational reform" (p. 341).

Research on teacher self-efficacy originated and flourished due to the powerful relationship between teacher beliefs, teacher change, and student achievement demonstrated in early studies (Armor et al, 1976). Since this time, the STEBI has helped science educators from many different countries gain a better understanding of elementary science teacher self-efficacy and the complex process of teacher education. This understanding can be deepened by continuing to refine the construct of self-efficacy, tending to instrumentation challenges and by using a variety of additional qualitative research methodologies to interpret or clarify teacher responses on quantitative instruments. Future research on science teacher self-efficacy also must continue to explore the relationships between self-efficacy beliefs, teaching practices, and student learning to move us closer to our vision of teaching and learning for a better future for all. As expressed 20 years ago, "Little will have been accomplished if research into educational beliefs fails to provide insights into the relationships between beliefs, on the one hand, and teacher practices, teacher knowledge, and student outcomes on the other" (Pajares, 1992, p. 327). We believe the STEBI can continue to play a role in this important work.

REFERENCES

Andersen, A. M., Dragsted, S., Evans, R. H., & Sorensen, H. (2003). Transforming the standard instrument for assessing science teacher's self-efficacy beliefs (STEBI) for use in Denmark. In D. Psillos, P. Kariotoglou, V. Tselfes, E. Hatzikraniotis, G. Fassoulopoulos, M. Kallery (Eds.), *Science Education Research in the Knowledge-Based Society*. Kluwer Academic Publishers: Dordrecht, Netherlands.

Armor et al. (1976). *Analysis of the school preferred reading programs in selected Los Angeles minority schools* (Rep. no. R-2007-LAUSD). Santa Monica, CA: RAND.

Ashton, P. T., & Webb, R. B. (1986). *Making a difference: Teachers" sense of efficacy and student achievement.* Santa Monica, CA, Longman, New York, NY.

Aydin, S., & Boz, Y. (2010). Pre-service elementary science teachers' science teaching efficacy beliefs and their sources. *Elementary Education Online, 9*(2), 694–704.

Azar, A. (2010). In-serice and pre-service secondary science teachers' self-efficacy beliefs about science teaching. *International Research and Reviews, 5*(4), 175–188.

Bandura, A. (1997). *Self-Efficacy. The exercise of control.* New York, NY : Freeman.

Bandura, A. (1977). Self-efficacy: Toward a unifying theory of behavioral change. *Psychological Review, 84*(2), 191–215.

Bandura, A. (1986). *Social foundations of thought and action: A social cognitive theory.* Prentice-Hall, Englewood Cliffs, NJ.

Berman, P., McLaughlin, M., Bass, G., Pauly, E., & Zellman, G. (1977). *Federal programs supporting educational change, Vol. 7: Factors affecting implementation and continuation.* (No. R-1589/7-HEW). Santa Monica, CA: Rand Questions.

Bleicher, R., E. (2006). Nurturing confidence in preservice elementary science teachers. *Journal of Science Teacher Education, 17*(2), 165–187.

Bleicher, R. (2004). Revisiting the STEBI-B: Measuring self-efficacy in preservice elementary teachers. *School Science and Mathematics, 104,* 383–391.

Bleicher, R. E., & Lindgren, J. (2005). Success in science learning and preservice science teaching self-efficacy. *Journal of Science Teacher Education, 16*(3), 205–225.

Boone, W. J., Townsend, J. S., & Staver, J. (2011). Using Rasch theory to guide the practice of survey development and survey data analysis in science education and to inform science reform efforts: An exemplar utilizing STEBI self-efficacy data. *Science Education, 95,* 258–280.

Bursal, M. (2008). Changes in Turkish pre-service elementary teachers' personal science teaching efficacy beliefs and science anxieties during a science methods course. *Turkish Science Education, 5*(1), 99–112.

Cakiroglu, J., Cakiroglu, E., & Boone, W. J. (2005). Pre-service teacher self-efficacy beliefs regarding science teaching: A comparison of pre-service teachers in Turkey and the USA. *Science Educator, 14,* 31–40.

Cannon, J. R., & Scharmann, L. C. (1995). Influence of cooperative early field experience on preservice elementary teachers' science self-efficacy. *Science Education, 80,* 419–436.

Cantrell, P., Young, S., & Moore, A. (2003). Factors affecting science teaching efficacy of preservice elementary teachers. *Journal of Science Teacher Education, 14*(3), 177–192.

Christol, P. G., & Adams, A. D. (2006). *Using the STEBI-B to determine the impacts of a standards-driven course on pre-service students' sense of personal and teaching efficacy in science education.* ASTE Conference Proceedings. Retrieved from http://theaste.org/publications/proceedings/2006proceedings/index.htm

Cone, N. (2009). A bridge to developing efficacious science teachers of ALL students: Community- based student learning supplemented with explicit discussions and activities about diversity. *Journal of Science Teacher Education, 20,* 365–383.

Czerniak, C. M., & Lumpe, A.T. (1996). Relationship between teacher beliefs and science Education reform. *Journal of Science Teacher Education, 7*(4), 247–266.

Dira-Smolleck, L. A. (2004). *The development and validation of an instrument to measure preservice teachers' self-efficacy in regard to the teaching of science as inquiry.* Unpublished doctoral dissertation, The Pennsylvania State University.

Enochs, L. G., & Riggs, I. M. (1990). Further development of an elementary science teaching efficacy belief instrument: A preservice elementary scale. *School Science and Mathematics, 90*(8), 694–712.

Enochs, L. G., Scharmann, L. C., & Riggs, I. M. (1995). The relationship of pupil control to elementary science teacher self-efficacy and outcome expectancy. *Science Education, 79*(1), 63–75.

Enochs, L. G., Smith, P. L., & Huinker, D. (2000). Establishing factorial validity of the mathematics teaching efficacy beliefs instrument. *School Science and Mathematics, 100*(4), 194–213.

Gibson, S., & Dembo, M. (1984). Teacher efficacy: A construct validation. *Journal of Educational Psychology, 76,* 569–582.

Ginns, I. S., & Watters, J. J. (1999). Beginning elementary school teachers and the effective teaching of science. *Journal of Science Teacher Education, 10*(4), 287–313.

Ginns, I. S., Watters, J. J., Tulip, D. F., & Lucas, K. G. (1995). Changes in preservice elementary teachers' sense of efficacy in teaching science. *School Science and Mathematics, 95*, 394–400.

Haney, J. J., Lumpe, A. T., Czerniak, C. M., & Egan, V. (2002). From beliefs to actions: The beliefs and actions of teachers implementing change. *Journal of Science Teacher Education, 13*(3), 171–187.

Henson, R. K. (2002). From adolescent angst to adulthood: Substantive implication and measurement dilemmas in the development of teacher efficacy research. *Educational Psychologist, 37*(3), 137–150.

Holden, M. E., Grouix, J., Bloom, M. A., & Weinburgh, M. H. (2011). Assessing teacher self-efficacy through an outdoor professional development experience. *Electronic Journal of Science Education, 12*(2),1–25.

Hoy, W. A. (2000). *Changes in teacher efficacy during the early years of teaching.* A paper presented at the Annual Meeting of the American Educational Research Association. New Orleans.

Labone, E. (2004). Teacher efficacy: Maturing the concept through research in alternative paradigms. *Teaching and Teacher Education, 20*, 341–359.

Lardy, C. H., & Mason, C. L. (2011, June). *Investigating reform and comparison courses: Long-term impact on elementary teachers' self-efficacy.* A paper presented at the NSEUS National Conference on Research Based Undergraduate Science Teaching: Investigating Reform in Classrooms, Bryant Conference Center, University of Alabama, Tuscaloosa, AL.

Liang, L. L., & Richardson, G. M. (2009). Enhancing prospective teachers' science teaching efficacy Beliefs through scaffolded, student-directed inquiry. *Journal of Elementary Science Education, 21*(1), 51–66.

Lindgren, J., & Bleicher, R. (2003, March). *Understanding and use of the learning cycle and elementary preservice teachers' self-efficacy.* Paper presented at the Annual Meeting of the National Association of Research in Science Teaching, Philadelphia.

Lloyd, J. K., Smith, R. G., Fay, C. L., Khang, G. N., Wah, L. L. K., & Sai, C. L. (1998). Subject knowledge for science teaching at primary level: A comparison of preservice teachers in England and Singapore. *International Journal of Science Education, 20*, 521–532.

McKinnon, M. C. (2010). *Influences on the science teaching self efficacy beliefs of Australian primary school teachers.* Retrieved from the ANU Digital Thesis Collection: http://hdl.handle.net/1885/9148)

Mihladiz, A. G., Duran, B. M., Isik, C. H., & Ozdemir, D. O. (2010). *The relationship between the pre-service science teachers' self-efficacy beliefs about science teaching and laboratory works.* Western Anatolia Journal of Educational Science, Special Issue: selected papers presented at WCNTSE.

Mulholland, J., Dorman, J. P., & Odgers, B. M. (2004). Assessment of science teaching efficacy of preservice teachers in an Australian University, *Journal of Science Teacher Education, 15*, 313–333.

National Research Council. (2000). *Inquiry and the national science education standards: A guide for teaching and learning.* Washington, DC: National Academy Press.

National Science Foundation. (1985) *Kansas outstanding rural science teacher project: A program of recognition now and preparing for the future.* Kansas State University. (NSF TEI 84-70338).

Palmer, D. (2006). Durability of changes in self-efficacy of preservice primary teachers. *International Journal of Science Education, 28*, 655–671.

Pajares, M. F. (1992). Teachers' beliefs and educational research: Cleaning up a messy construct, *Review of Educational Research, 62*(3), 307–332.

Perkins, C. J. (2007). *Toward making the invisible visible: Studying science teaching self-efficacy beliefs.* An unpublished dissertation, Oregon State University.

Ramey-Gassert, L., & Shroyer, M. G. (1992). Enhancing science teaching self-efficacy in preservice elementary teachers. *Journal of Elementary Science Education, 4*(1), 26–34.

Ramey-Gassert, L., Shroyer, M. G., & Staver, J. R. (1996). A qualitative study of factors influencing science teaching self-efficacy of elementary level teachers. *Science Education, 80(3)*, 283–315.

Richardson, G. M., & Liang, L. L. (2008). The use of inquiry in the development of preservice teacher efficacy in math and science. *Journal of Elementary Science Education, 20*(1), 1–16.

Ritter, J. M., Boone, W. J., & Rubba, P. A. (2001). Development of an instrument to assess prospective elementary teacher self-efficacy beliefs about equitable science teaching and learning (SEBEST). *Journal of Science Teacher Education, 12*, 175–198.

Riggs, I. M., & Enochs, L. G. (1990). Toward the development of an elementary teacher's science teaching efficacy belief instrument. *Science Education, 74*(6), 625–637.

Riggs, I. M., Enochs, L. G., & Posnanski, T. J. (1998). *The teaching behaviors of high versus low efficacy elementary teachers.* A paper presented at the Annual' Meeting of the National Association for Research in Science Teaching, San Diego, CA.

Rotter, J. B. (1966). Generalized expectancies for internal vs. external control of reinforcement. *Psychological Monographs, 80,* 1–28.

Rubeck, M., & Enochs, L. G. (1991). *A path analytic model of variables that influence science and chemistry teaching self-efficacy and outcome expectancy in middleschool science teachers.* A paper presented at the Annual Meeting of the National Association for Research in Science Teaching. Lake Geneva, WI.

Scharmann, L. C., & Hampton, C. M. O. (1995). Cooperative learning and pre-service elementary teacher science self-efficacy. *Journal of Science Teacher Education, 6,* 125–133.

Schoon, K. J., & Boone, W. J. (1998). Self-efficacy and alternative conceptions of science of preservice elementary teachers. *Science Education, 82*(5), 553–568.

Settlage, J. (2000). Understanding the learning cycle: Influences on abilities to embrace the approach by preservice elementary school teachers. *Science Education, 84,* 43–50.

Settlage, J., Southerland, S. A., Smith, L. K., & Ceglie, R. (2009). Constructing a doubt-free teaching self: Self-efficacy, teacher identity, and science instruction within diverse settings, *Journal of Research in Science Teaching, 46*(1), 102–125.

Shroyer, M. G. (1997). *Enhancing preservice elementary teachers science teaching self-efficacy beliefs: A longitudinal inquiry.* A paper presented at the Annual Meeting of the Association for the Education of Teachers of Science. Cincinnati, OH.

Smolleck, L. D., Zembal-Saul, C., & Yoder, E. P. (2006). The development and validation of an instrument to measure pre-service teachers' self-efficacy in regard to the teaching of science as inquiry. *Journal of Science Teacher Education, 17*(2), 137–163.

Stevens, C., & Wenner, G. (1996). Elementary preservice teachers' knowledge and beliefs regarding science and mathematics. *School Science and Mathematics, 96*(1), 2–9.

Tschannen-Moran, M., Hoy, A. W., & Hoy, W. K. (1998). Teacher efficacy: Its meaning and measure. *Review of Educational Research, 68*(2), 202–248.

Tosun, T. (2000). The impact of prior science course experience and achievement on the science teaching self efficacy of preservice elementary teachers. *Journal of Elementary Science Education, 12*(2), 21–31.

Watters, J. J., & Ginns, I. S. (2000). Developing motivation to teach elementary science: Effect of collaborative and authentic learning practice in preservice education. *Journal of Science Teacher Education, 11,* 301–321.

Webb, P., & Gripper, A. (2010). Developing teacher self-efficacy via a formal HIV/AIDS intervention. *Journal of Social Aspects of HIV/AIDS, 7*(3), 28–34.

WEE-LOON, NG. (2011). *A study of Singapore female primary teachers' self-efficacy for teaching science.* An unpublished dissertation, Durham University.

Wheatley, K. F. (2005). The case for re-conceptualizing teacher efficacy research. *Teaching and Teacher Education, 21,* 747–766.

Wilson, J. (1996). An evaluation of the field experiences of the innovative model for the preparation of elementary teachers for science, mathematics, and technology. *Journal of Teacher Education, 47*(1), 53–59.

Yilmaz, H., & Cavas, P. H. (2008). The effect of the teaching practice on pre-service elementary science teachers' science teaching efficacy and classroom management beliefs. *Eurasia Journal of Mathematics, Science, & Technology, 4*(1), 45–54.

Young, B. J., & Kellogg, T. (1993). Science attitudes and preparation of preservice elementary teachers. *Science Education, 77*(3), 279–291.

AFFILIATIONS

Gail Shroyer
Department of Curriculum and Instruction
College of Education
Kansas State University

Iris M. Riggs
Department of Science, Mathematics, and Technology Education
College of Education
California State University, San Bernardino

Larry G Enochs
Professor Emeritus
Science and Mathematics Education
Oregon State University

SECTION THREE

POLICY AND BELIEFS RESEARCH

CELESTINE H. PEA

NSF-FUNDED RESEARCH ON BELEIFS
IN STEM EDUCATION

This material is based upon work supported by (while serving at) the National Science Foundation. "Any opinions, findings, and conclusions or recommendations expressed in this chapter are those of the author and do not reflect the views of the National Science Foundation in any way."

INTRODUCTION

This chapter looks at nearly two decades of research about beliefs on topics in science, technology, engineering, and mathematics (STEM) fields and science education funded by the National Science Foundation (NSF). These studies encompass nearly every educational level and key policy recommendation in both formal and informal settings. The population that comprises these studies includes college and universities faculties and students, pre-service and in-service teachers, K-12 students and children, and policy makers. The research resides within the various scientific disciplines and in the areas of cognitive and social sciences, psychology, social justice, and policy. While the research featured in this chapter was funded by various offices and directorates across NSF, this chapter focuses almost exclusively on a subset of funded research in science education with some emphasis on the disciplines of physics, chemistry, biology, and engineering.

NSF operates under the auspices of the National Science Board (NSB) that serves as a national policy advisor to the President and the Congress and as the governing body for NSF. In this role, the Foundation supports a portfolio of investments that reflect the interdependence among fields, promote disciplinary strength, and embrace interdisciplinary activities. During fiscal year 2014, NSF funds more than 343,000 researchers, postdoctoral fellows, graduate assistant, trainees, teachers, and students, who actively engage or participate in a full range of research studies.

As the only federal agency dedicated to supporting basic research in all scientific fields and science education in the United States of America, NSF has the responsibility for promoting the progress of science, advancing the national health, prosperity, and welfare, and for engaging in other scientific-related activities aimed at expanding the boundaries of knowledge nationally and internationally. NSF does not operate laboratories of its own. Instead, the Foundation brings together diverse elements of the larger STEM communities to achieve its mission. Therefore,

R. Evans et al., (Eds.), The Role of Science Teachers' Beliefs in International Classrooms, 121–133.
© 2014 Sense Publishers. All rights reserved.

academic institutions in collaboration with public and private sectors are critical to NSF's support for catalyzing emerging opportunities in research and education (NSB, 2000).

Beyond the borders of the United States, the Foundation engages in strategic collaborations by policy that includes provisions for promoting international STEM research. NSF operates oversea offices in Paris, France; Tokyo, Japan; and Beijing, China, to foster jointly developed basic research projects. International projects help scientists and educators from different countries work collaboratively to advance STEM research. In addition, international collaborations occur through the Office of International and Integrative Activities that serves as the focal point for NSF's collaborative activities.

Special international collaborations, such as Science Across Virtual Institutes, provide a platform for teams of NSF-funded United States scientists and engineers to collaborate with their international counterparts. The intent of these collaborations is to enhance research studies, data sharing, networking, and technical exchanges among students, postdoctoral researchers, and junior faculty whose strengths and interests complement each other.

HIGHER EDUCATION AND TEACHER EDUCATION

Using National Science Board (NSB, 2010; 2007) reports as a backdrop, it seems reasonable and appropriate that capturing the essence of research on beliefs in science education is relevant to national policy research agendas since beliefs are the best indicators of decisions individuals make throughout their lives (Fishbein & Ajzen, 1975; Pajares, 1992; Rokeach, 1968). In a given situation, beliefs form attitudes, which lead to decisions and behaviors that influence people's actions—including K-20 classrooms and beyond (Pajares, 1992). These actions provide insights into the perspectives people hold about teaching and learning and for broadening participation of all students in STEM fields. NSB (2010) policy recommendations aim to help the Nation prepare a greater number of students to succeed in the 21st Century. NSF-funded studies in science education focus on both theory building and theory testing in order to advance the knowledge base about innovative, emerging, and best practices regarding beliefs in science education. Results from these studies can provide policy makers with evidence to guide decisions about teaching and learning and broadening participation in the United States educational and abroad.

Higher Education Studies

Many of the new ideas, strategies, models, and interventions that emerge from NSF support, funded the development of a more scientifically literate citizenry and STEM workforce (Fullan, 2007; PCAST, 2012; NSB, 2000; Spillane, 2012). In the area of beliefs, most studies examine the connection of beliefs with instructional practices and the rigor of course offerings. For example, Henderson, Yerushalmi, Kuo,

Heller, and Heller (2007); Duran, Czerniak, and Haney (2005); and Southerland, Gess-Newsome, and Johnston (2003) investigated scientists' beliefs about their instructional practices, what shape university faculties' selection and use of various teaching methods, and the actual use of strategies that encourage greater student participation in STEM.

Other studies help faculty members, postdoctoral professionals, and graduate and undergraduate students let go of entrenched beliefs about less effective instructional methods in favor of more successful ones (e.g., Goertzen, Scherr, & Elby, 2010; McGinnis, et al., 2002). To facilitate this process, studies investigate the beliefs these professionals bring to the teaching and learning environment. Still other studies explore better ways to handle, for instance, the cultural norms, time pressures, or reward systems in higher education that interfere with broad implementation of more effective pedagogical approaches at the collegiate level. All of these studies show potential for advancing knowledge in the field as described in NSB's research and policy recommendations (NSB, 2010; 2007).

Teacher Education

Scores of rigorous studies, reports commissioned by national policy makers, and student achievement scores on national and international tests show that the development of a STEM-capable citizenry, proficient workforce, or future experts are not occurring at a level necessary to help keep the United States globally competitive in STEM (NSB, 2010; PCAST, 2012). As information from these sources became more commonly known, NSF continued to provide support at every level in science education to help ensure equity through excellence for all students. These efforts include an increased emphasis on understanding the needs of and enhancing the professional growth of both K-12 pre-service and in-service teachers.

To help the Nation reach this laudable goal, researchers examine a range of factors related to the beliefs teachers hold. Some studies describe the beliefs of teachers in reference to their own process of learning to teach, while others look at the contextual environment in which teachers teach. Researchers also focus on teachers' beliefs in conjunction with their personal histories (Eick & Reed, 2002), the influence of their beliefs on the development of their pedagogical content knowledge and instructional strategies (Brickhouse, Bodner, & Neie, 1987), and the weight of the classroom context on the beliefs of teachers (Hancock & Gallard, 2004). Additional research explores teachers' beliefs relative to the induction or preparation period (Fletcher & Luft, 2011; Hancock & Gallard, 2004; Luft et al., 2011; Luft, Roehrig, & Patterson, 2003; Marbach & McGinnis, 2008). Collectively, these studies reveal the complex process of becoming a teacher and reinforce the importance of having a highly qualified K-12 teaching workforce to improve student learning in STEM fields.

Change. To ensure that the science teaching workforce is capable of preparing students to participate fully in the 21ˢᵗ Century STEM enterprise, research is carried out on factors other than those related primarily to teacher preparation and instruction.

Studies investigate ways to prepare teachers for the challenges they encounter once they are in their own classrooms. Researchers who explore teacher beliefs in this area suggest that recognizing the need for and accepting change associated with new policy mandates is one of the biggest challenges teachers face in the classroom. Change can be difficult for teachers because new policies may conflict with their existing beliefs (Haney, Czerniak, & Lumpe, 1996; Yerrick, Park, & Nugent, 1997). In addition, new and more innovative teaching strategies and changing school and classroom environments may also be at odds with the beliefs teachers hold about teaching and learning and may limit the implementation of changes in science education policies (Davis, 2002; Haney & McArthur, 2002; Haney, Lumpe, & Czerniak, & Egan, 2002). Overall, this misalignment of teacher beliefs and policy mandates often result in limited teacher change.

Researchers, however, are finding that providing opportunities for teachers to personally experience desired changes are critical to the adoption of newer reform-based practices. When teachers experience the outcome of the policy (e.g., inquiry instruction, scientific practices) and they participate in its implementation (e.g., professional learning communities), they are more likely to adopt the reform or the policy (Davis, 2002; Marbach & McGinnis, 2008; Roehrig, Kruse, & Kern, 2007). Feldman (2000) suggests that models are needed to better understand the connection of teachers' beliefs, reasoning, and knowledge regarding change involved in science education. With models, programs can be developed to support teacher change. Without such models, changes in teachers' beliefs become more unlikely, regardless of the benefits the change might bring to the teachers and possible student learning (Feldman, 2000; Wallace & Kang, 2004).

Assessment. Research on beliefs about teaching and learning would not be complete without studies about how best to measure teacher beliefs. NSF has a long history of supporting various assessment efforts through national educational agencies, leading testing entities, public policy and research consulting firms and colleges and universities. One university-based study by Lumpe, Haney, and Czerniak (2000) resulted in the development of an instrument to assess teachers' beliefs during and following their participation in a long-term professional development program. Another study involved the synthesis of existing assessment measures, which led to the release of the *Compendium of Research Instruments for STEM Education: Part 1* (Minner, Martinez, & Freeman, 2012, August) and the *Compendium of Research Instruments for STEM Education: Part 2* (Minner, Ericson, Wu, & Martine, 2012, November). These documents highlight nine instruments dedicated to measuring instructional beliefs of teachers; six of them focus on science.

SOCIAL AND PSYCHOLOGICAL FACTORS

Studies about science education that focus on scientific principles and practices associated with science content, instruction, and assessment portray one side of teaching and learning. Studies also need to be conducted on factors that might influence more students to select, excel, and remain in STEM careers. Such studies often fall in the social and psychological domains.

Culture and Diversity

Culture. To provide all students with an ecosystem that helps them reach their fullest potential, activities and practices must expansively enlist culture and diversity as a creative source in STEM teaching and learning. NSF fully embeds culture and diversity in its efforts to develop a world-class STEM workforce that promotes equity that embodies excellence in education. Since schools and classrooms are gathering points for different cultures of students, and since researchers are looking for ways to develop models and practices that reflect culture relevance, researchers nest studies in culture-rich community settings to further uncover the value culture adds to the learning environment (Bang, Medin, & Atran, 2007; Lin &, Schwartz, 2003).

Confronted with the need to understand how culture influence student learning, Medin, Waxman, Woodring, and Washinawatok (2010) conducted an interdisciplinary study to examine how underrepresented minorities (URM) students learn fundamental science content. Results show that research should attend to the different forms of culture (e.g. linguistic, epistemological, or societal) when attempting to understand how students learn science. The results also help to dispel the myth that improving academic performance in STEM fields and increasing URM participation in the future STEM workforce rest exclusively on classroom learning restricted largely to content, instruction, and assessment. Instead, studies that relate, connect, and integrate factors across educational and socio-psychological domains also help increase the number of URMs who excel in STEM careers (Kaiser, 2011; NRC, 2012).

Washinawatok (1993), also examined the role of culture in the development of knowledge and reasoning for students whose self-esteem was lowered when their native language and culture were removed from the school curriculum. Washinawatok's research show reversing such actions restores students' beliefs about the value of schooling, increases their academic skills, and enhances interactions with others. Accordingly, future theories aimed at increasing URM's participation in the scientific workforce must comprehensively accommodate a wide range of community characteristics than reach far beyond methodological consistency or representative sampling. Rather, it involves a direct appeal for researchers to provide evidence-based models of the valuable contributions that diverse cultures bring to the teaching and learning process in STEM fields (Medin et al.,2010; NSB 2007).

Diversity. Studies about diversity span the full gamut of possibilities, including supporting positive teacher beliefs about the increasing diversity of today's classroom, changing teacher beliefs as a result of interventions that increase diversity awareness, and providing professional learning opportunities that encourage teachers to value the diversity within their learning environments. These studies recognize the importance of the diverse theories both teachers and students bring to the classroom about science learning (Eccles & Wigfield, 2002; Lin & Schwarz, 2007), and explore ways to help both teachers and students come to a better understanding of how these theories might unfold in the learning environment.

To address some of the challenges diversity might bring in the classroom, Lee (2004) looked at linguistic issues and examined how elementary Hispanic teachers change their beliefs and practices. The focus was on how well these teachers establish internal congruence through the process of mediating academic disciplines with linguistic and cultural experiences of diverse student groups. Results suggest that establishing congruence is gradual and demanding and requires ongoing formal training, scaffolding support, and collaborative sharing for teachers to be engage skilfully in practices that matters most.

From a different viewpoint, Lin and Schwarz (2007) investigated how teachers need to become more adaptive experts to successfully address the value diverse students bring to the classroom. To that end, these researchers explored the value theories both teachers and students bring to the classroom around science learning to help teachers develop habits of seeking relevant pedagogical teaching strategies that best match the diversity of their students. Lin and Schwartz (2007) combined socio-cultural, cognitive and instructional approaches in a study to test the claim that centralizing teachers' and students' values helps teachers adapt their instruction to the diversity of students. The overarching goal was to reduce mismatches between teachers and students beliefs about learning science by looking beyond diversity as a demographic variable. Lin and Schwartz sought to include diversity as a variable that brought values to innovations and discoveries in science. This research led to the development of a new interdisciplinary center that serves as the core site where adaptive expertise is built and adaptations among people and ideas are commonplace.

Motivation and Interest

Motivation. Motivation is a critical to students' success in STEM careers. Researchers (Bang & Medin, 2010; Koballa & Glynn, 2007) studied the importance of motivation in one's success, and suggest that skills and the availability of supportive learning environments are important to this success. Similar research confirms that when support is lost, beliefs systems and motivation can become easily compromised (Bang & Medin, 2010; Eccles & Wigfield, 2002; Simpkins, Davis-Kean, & Eccles, 2006) and students lose the desire to keep trying.

For URM and women, if conditions are not favorable, their beliefs systems can be weakened, which sometimes negatively impacts their primary and subsequent choices about STEM careers (Hazari, Sonnert, Sadler, & Shanahan, 2010, Nosek et al., 2009; Wigfield & Eccles, 2000). Without URM and women's consideration of STEM careers, policy mandates that call for increased numbers in the STEM workforce (e.g., NRC, 2012; NSB, 2010; PCAST, 2012) will not likely be met.

Interest. Like motivation, students' interest in STEM is being studied on every education level and across multiple fields. Areas of concern among policymakers, educators, and researchers pertain to the loss of interest in STEM subjects by students in the early grades and by students who excel in STEM disciplines, yet somehow lose interest over time and drop out of the STEM pipeline (PCAST, 2012), forgo careers in STEM, or eventually leave STEM jobs (Krapp & Prenzel, 2011). Policy makers (PCAST, 2010; 2012) and researchers agree that to meet the future STEM workforce all of these concerns must be addressed aggressively.

With mounting evidence of the loss of interest in STEM fields and careers by students at all levels and all ages (PCAST, 2010), research shows that the loss of interest is directly linked to the teaching and learning environment, student-teacher interactions, instructional practices, and difficulty with STEM content. One major national effort underway to respond to these findings is enlisting the broader community to retain students in the STEM pipeline and in their STEM careers (PCAST, 2010; 2012). To also help reverse this trend and to address these concerns, Hullenman & Harackiewicz (2009) used a motivational intervention to investigate whether relevance of science to students' lives promote interest and performance. Results suggest that encouraging students to make connections between science course materials and their lives promote both interest and performance for students. The effect was most striking in low-expectancies students who improved nearly two-third of a letter grade in the relevance condition of the intervention. Other studies investigated whether enrolment, persistence, or shared interested had anything to do with the lack of interest and confidence in STEM by URM students. One such study was conducted by Sweeder, Strong, Shipman, and Jonelle (2009) who monitored a science scholarship program to test if students with a shared interest in science benefit from living and working together. Results show that students benefitted from the specialized program, which helped them to discover and reaffirm their beliefs about their career decisions and become more self-confident and empowered about what they have the ability to do.

Stereotype Threats and Self-affirmation Interventions

There is general agreement that situational environmental cues interfere with the participation of individuals in STEM fields. NSB (2010) notes that:

> Intellectually talented children and young adults can readily detect ambivalence, low expectation, or other negative attitudes within their learning ecosystems. Worse yet, sometimes these student face outright hostility. This often results in adverse consequences, such as poor self-efficacy, loss of motivation, and intellectual regression. (p 22).

For many researchers (Aronson, 2002; Aronson, Fried, & Good, 2002; Cohen & Garcia, 2008), addressing these issues are important in identifying and eliminating barriers to participation in STEM careers brought on by negative situational cues frequently associated with stereotype threats. These threats, which can include assumptions about learning and performance, can hinder students' ability to excel in

STEM fields despite strong personal agency beliefs, self-efficacy beliefs, and high academic ability.

Fearing that students might not be able to handle such barriers, researchers (Cohen & Garcia, 2008; Purdie-Vaughns et al., 2009) investigated ways to address these threats in terms of students' self-beliefs, self-efficacy, and self-concept beliefs. These researchers developed self-affirmation interventions to help students break recursive cycles or adjust to changes in their learning environment that interfere with their ability to succeed (Cohen, Garcia, Apfel & Master, 2006; Kurtz-Costes & Rowley, 2012; Miyake et al., 2010; Purdie-Vaughns et al., 2009; Sherman & Cohen, 2006; Sherman et al., 2013).

Cohen and Garcia (2008) also developed the Identity Engagement Model of the effects of social threat on academic performance. The model outlines a series of actions students can take to ameliorate the threatening conditions, confirm or disconfirm if an actual threat exists, set in motion an appraisal process about how to deal with the threat, and assess whether they have the ability and/or desire to cope with the threat (Cohen & Garcia, 2008; Cohen, Garcia, Purdie-Vaughns, Apfel, & Brzustoski, 2009). Despite these advances in helping students deal with stereotype threats, more research is needed to study the influence of self-perceptions, cultural identity, and stereotypes threats on student learning and broadening participation in STEM. Additional research is also needed to develop more interventions and to test existing ones in multiple contexts.

Gendered Differences: Creating a Sense of Belonging

Increasingly, education policy mandates in the United States are addressing inequities associated with the recruitment and retention of women in the hard sciences (e.g., engineering, physics, computer science) and science education (Cheryan, Plaut, Davies, & Steel, 2009; Eccles, 2011; Hill, Corbett, & St. Rose, 2010). In particular, there has been an upwelling of attention to individual and structural barriers that block women from full participation in collaborative research opportunities, mentoring activities, professional relationships associated with work-related growth and upward social mobility due primarily to their gender (Ko, Kachchaf, Ong, & Hodari, 2013; NSB, 2012; 2000).

Research suggests that negative influences on women in STEM fields occur at both the individual and structural levels. To improve conditions in these areas, Fox, Sonnert, and Nikiforova (2011) studied programs for undergraduate women in science and engineering at a major research university to showcase disparities between males and females within higher education. Their study showed that structural features, not individuals as the major causes of issues for women. While people who implement activities to address problems women face, typically focus on the structural level with training in diversity or mentoring, Fox et al. (2011) learned that this approach is often misaligned with the real problems women face.

Considering this type of research as an important topic for deeper study, Cheryan, Meltzoff, and Kim (2011) implemented an intervention whose outcomes call for both structural and environmental changes. These recommendations set in motion actions that caused policy-level authorities to change their behaviors if not their beliefs. This change led to environmental changes that promote a sense of belonging and a broad appeal to women in computer science (Cheryan et al., 2011; Cohen & Garcia, 2008). As a result of these changes, trend data show women's percent of bachelors' degrees increased considerably.

Research shows that women are overcoming these barriers (Cheryan et al., 2011; Fox et al., 2011), yet workable solutions to reduce such barriers are far from complete. In the meantime, data show that women frequently lessen the influence of stress brought on by STEM-related careers though proactive networks, temporary absences from STEM environments, hobbies, and participation in career life-balance initiatives. In 2011, NSF launched a career life-balance program to bolster development of the STEM workforce (see www.nsf.gov/ Career-life-balance).

SUMMARY

Beliefs are shaped by experiences from direct and indirect interactions (e.g., newspapers, books, magazines, radio and television, lecturers, friends, relatives, coworkers, Twitter, Facebook) throughout our lives. Whatever beliefs turn out to be, they influence an individual's actions and behaviors. Since beliefs are vast, highly personal, and not subject to persuasion, researchers look to studies of people's actions and behaviors for insights about how these beliefs play out in the teaching and learning environment. Researchers wish to better understand the beliefs people hold, which are sometimes entrenched and often conflict with policy mandates and recommendations, adoption of new curriculums and teaching pedagogies, and/or proposed changes in the teaching and learning environment. When such conflicts occur, fewer students do well in STEM, frequently lose interest in these areas, sometimes switch out of STEM majors, and leave jobs in STEM fields.

Whereas these conditions are evident in the education system of the United States, results from disciplinary and interdisciplinary studies within and across cognitive and affective domains are informing and binding together a stronger foundation for keeping the Nation competitive worldwide. NSF-funded research is linking scientific research and education in meaningful ways both nationally and internationally. Support for this claim is evident in advances on scientific and technological forefronts, meaningful education exchanges, international co-authorships, and the number of students who study abroad in the United States. The support by NSF is also evident in rich research findings that contain innovative ideas, models, and tools; better frameworks and protocols; newer platforms and networks; and improved research methods and metrics. All of these outcomes are essential to transforming science education through basic research; particularly in the areas of learning and broadening participation in science through a stronger and well-informed STEM education system and workforce.

REFERENCES

Aronson, J. (2002). Stereotype threat: Contending and coping with unnerving expectations. In J. Aronson (Ed.), *Improving academic achievement: Impact of psychological factors on education* (pp. 303–328). San Diego, CA: Academic Press.

Aronson, J., Fried, C. B., & Good, C. (2002). Reducing the effects of stereotype threat on African American college students by shaping theories of intelligence. *Journal of Experimental Social Psychology, 38*, 113–125.

Bang, M., & Medin, D. (2010). Cultural processes in science education: Supporting the navigation of multiple epistemologies. *Science Education, 94*(6), 1008–1026. doi:10.1002/sce.20392.

Bang, M., Medin, D. L., & Atran, S. (2007). Cultural mosaics and mental models of nature. *Proceedings of the National Academy of Sciences, 104*(35), 13868–13874.

Brickhouse, N. W., Bodner, G. M., & Neie, V. N. (1987). Teacher beliefs about science and their influence on classroom practice. In Proceedings from the Second International Seminar of Misconceptions and Educational Strategies in Science and Mathematics 2, 34-48.

Cheryan, S., Meltzoff, A. N., & Kim, S. (2011). Classroom matter: The design of virtual classrooms influences gender disparities in computer science classes. *Computers & Education, 57*(2), 1825–1835.

Cheryan, S., Plaut, V. C., Davies, P. G., & Steele, C. M. (2009). Ambient belonging: how stereotypical cues impact gender participation in computer science. *Journal of Personality and Social Psychology, 97*(6), 1045.

Cohen, G. L., & Garcia, J. (2008). Identity, belonging, and achievement: A model, interventions, implications. *Psychological Science, 17*(6), 365–369.

Cohen, G. L., Garcia, J., Apfel, N., & Master, A. (2006). Reducing the racial achievement gap: A social-psychological intervention. *Science, 313*(5791), 1307–1310.

Cohen, G. L., Garcia, J., Purdie-Vaughns, V., Apfel, N., & Brzustoski, P. (2009). Recursive processes in self-affirmation: Intervening to close the minority achievement gap. *Science, 324*(5925), 400–403.

Davis, K. (2002). Change is hard: What science teachers are telling us about reform and teacher learning of innovative practices. *Science Education, 87*(1), 3–30.

Duran, L. P, Czerniak, C. M., & Haney, J. (2005). A study of the effects of a LSC project on scientists' teaching practices and beliefs. *Journal of Science Teacher Education, 16*(2), 159–184.

Duschl, R. A., Schweingruber, H. A., & Shouse, A. W. (Eds.). (2007). *Taking science to school: Learning and teaching science in grades K-8.* National Academies Press.

Eccles, J. S. (2011). Gendered educational and occupational choices: Applying the Eccles et al. model of achievement-related choices. *International Journal of Behavior Development, 35*(3), 195–201. doi: 10.1177/0165025411398185

Eccles, J. S., & Wigfield, A. (2002). Motivational beliefs, values, and goals. *Annual Review of Psychology, 53*(1), 109–132.

Eick, C. J., & Reed, C. J. (2002). What makes an inquiry-oriented science teacher? The influence of learning histories on student teacher role identify and practice. *Science Education, 86*(3), 401–416.

Feldman, A. (2000). Decision-making in the practical domain: A model of practical conceptual change. *Science Education, 84*(5), 606–623

Fishbein, M., & Ajzen, I. (1975). *Beliefs, attitude, intention, and behavior: An introduction to theory and research.* Reading, MA: Addison-Wesley.

Fletcher, S. S., & Luft, J. A. (2011). Early career secondary science teachers: A longitudinal study of beliefs in relation to field experiences. *Science Education, 95*(6), 1124–1146.

Fox, M. F., Sonnert, G., & Nikiforova, I. (2011). Programs for undergraduate women in science and engineering: Issues, problems and solutions. *Gender & Society, 25*(5), 589–615. doi:10.1177/0891243211416809

Fullan, M. G. (2007). Change the term for teacher learning. *Journal of Staff Development, 28*(3), 35–36.

Goertzen, R. M., Scherr, R. E., & Elby, A. (2010). Tutorial teaching assistants in the classroom: Similar teaching behaviors are supported by varied beliefs about teaching and learning. *Physical Review Special Topics-Physics Education Research, 6*(1), 1–7. doi:10.1103/PhysRevSTPER.6.010105

Hancock, E. S., & Gallard, A. G. (2004). Per-service science teachers' beliefs about teaching and learning: The influence of K-12 field experience. *Journal of Science Teacher Education, 15*(4), 281–291.

Haney, J. J., & McArthur, J. (2002). Four case studies of prospective science teachers' beliefs concerning constructivist teaching practices. *Science Education, 86*(6), 783–802. doi:10.1002/sce.10038

Haney, J. J., Lumpe, A. T., Czerniak, C. M., & Egan, V. (2002). From beliefs to actions: The beliefs and actions of teachers implementing change. *Journal of Science Teacher Education, 13*(3), 171–187.

Haney, J. J., Czerniak, C. M., & Lumpe, A. T. (1996). Teacher beliefs and intentions regarding the implementation of science education reform strands. *Journal of Research in Science Teaching, 33*(9), 971–993.

Hazari, Z., Sonnert, G., Sadler, P. M., & Shanahan, M. C. (2010). Connecting high school physics experiences, outcome expectations, physics identity, and physics career choice: A gender study. *Journal of Research in Science Teaching, 47*(8), 978–1003. doi:10.1002/tea.20363

Henderson, C., Yerushalmi, E., Kuo, V. H., Heller, K., & Heller, P. (2007). Physics faculty beliefs and values about the teaching and learning of problem solving. II. Procedures for measurement and analysis. *Physical Review Special Topics, Physics Education Research, 3*(2), 1–12. doi:10.1103/PhysRevSTPER.3.020110

Hill, C., Corbett, C., & St. Rose, A. (2010). *Why so few? Women in science, technology, engineering, and mathematics.* American Association of University Women. Washington, DC 6.

Hulleman, C. S., & Harackiewicz, J. M. (2009). Promoting interest and performance in high school science classes. *Science, 326*(5958), 1410–1412.

Ingersoll, R. M. (2011). Do we produce enough mathematics and science teachers? *Phi Delta Kappan, 92*(6), 37–41.

Ko, L. T., Kachchaf, R. R., Ong, M., & Hodari, A. K. (2013, January). *Narratives of the double bind: Intersectionality in life stories of women of color in physics, astrophysics and astronomy* (Vol. 1513, p. 222). AIP Conference Proceedings.

Koballa, T., & Glynn, S. (2007). Attitudinal and motivational constructs in science learning. In S. Abell & N. Lederman (Eds). *Handbook of Research on Science Education* (pp. 75–102). Mahwah NJ: Lawrence Erlbaum Associates.

Krapp, A., & Prenzel, M. (2011). Research on interest in science: Theories, methods, and findings. *International Journal of Science Education, 33*(1), 27–50.

Kurtz-Costes, B., & Rowley, S. J. (2012). School transitions and African American youth. In S. A. Karabenick & T. C. Urdan (Eds.), *Transitions across schools and cultures: Vol. 17, Advances in motivation and achievement* (pp. 27–54). doi:10.1108/S0749-7423(2012)0000017005

Lee, O., Luykx, A., Buxton, C., & Shaver, A. (2007). The challenge of altering elementary school teachers' beliefs and practices regarding linguistic and cultural diversity in science education. *Journal of Research in Science Teaching, 44*(9), 1269–1291.

Lin, X. D., & Schwartz, D. L. (2003). Reflection at the crossroad of cultures. *Mind, Culture & Activity, 10*(1), 9–25.

Luft, J. A., Firestone, J. B., Wong, S. S., Ortega, I., Adams, K., & Bang, E. (2011). Beginning secondary science teacher induction: A two, year mixed methods study. *Journal of Research in Science Teaching, 48*(10), 1199–1234.

Luft, J. A., Roehrig, G. H., & Patterson, N. C. (2003). Contrasting landscapes: A comparison of the impact of different induction programs on beginning secondary science teachers' practices, beliefs, and experiences. *Journal of Research in Science Teaching, 40*(1), 77–97.

Lumpe, A. T., Haney, J. J., & Czerniak, C. M. (2000). Assessing teachers' beliefs about their science teaching context. *Journal of Research in Science Teaching, 37*(3), 275–292.

Marbach-Ad, G., & McGinnis, J. R. (2008). To what extent do reform-prepared upper elementary and middle school science teachers maintain their beliefs and intended instructional actions as they are inducted into schools? *Journal of Science Teacher Education, 19*(2), 157–182.

McGinnis, J. R., Kramer, S., Shama, G., Graeber, A. O., Parker, C. A., & Watanabe, T. (2002). Undergraduates' attitudes and beliefs about subject matter and pedagogy measured periodically in a reform, based mathematics and science teacher preparation program. *Journal of Research in Science Teaching, 39*(8), 713–737.

Medin, D. S. (2013). *EHR distinguished lecture series: Culture and thought: Native American relational epistemologies.* National Science Foundation. Arlington, VA

Medin, D., Waxman, S., Woodring, J., & Washinawatok, K. (2010). Human-centeredness is not an universal feature of young children's reasoning: Culture and experience matter when reasoning about biological entities. *Cognitive Development, 25*(3), 197–207. doi:10.1016/j.cogdev.2010.02.001

Minner, D., Martinez, A., & Freeman, B. (2012, May). *Compendium of research instruments for STEM education. Part 1: Addendum added. Teacher practices, PCK, and content knowledge.* Cambridge, MA: Abt Associates. Retrieved from http://cadrek12.org/sites/default/files/Compendium%20of%20 STEM%20instruments%20Part%201.pdf

Minner, D., Ericson, E., Wu, S., Martinez, A. (2012, November). *Compendium of research instruments for STEM education. Part 2: Measuring students' knowledge, reasoning skills, and psychological attributes. Abt Associates.* Washington, DC. Retrieved from: http://abtassociates.com/Reports/2012/ Compendium-of-Research-Instruments-for-STEM-Ed- (1).aspx.

Miyake, A., Kost-Smith, L. E., Finkelstein, N. D., Pollock, S. J., Cohen, G. L., & Ito, T. A. (2010). Reducing the gender achievement gap in college science: A classroom study of values affirmation. *Science, 330*(6008), 1234–1237.

National Research Council. (2012). Education for Life and Work: Developing transferable knowledge and skills in the 21st Century. Committee on defining deeper learning and 21st century skills. J. W. Pellegrino & M. L. Hilton (Eds.), *Board on testing and assessment and board on science education, division of behavioral and social sciences and education.* Washington, DC: National Academies Press.

National Science Board. (2012). *Science & Engineering Indicators - 2012.* Arlington, VA: National Science Foundation, 2012 (NSB-12-01).

Nationa Science Board. (2010). *Preparing the next generation of STEM innovators: Identifying and developing our nation's human capital.* National Science Foundation. (NSB-05-10).

National Science Board. (2007). *A national plan for addressing the critical needs of the U.S. science, technology, engineering, and mathematics education system.* National Science Foundation. (NSB-10-07).

National Science Board. (2000). *Science & engineering indicators - 2000.* Arlington, VA: National Science Foundation, 2000 (NSB-00-00).

National Science Foundation. (2001). *Balancing the scale: NSF's Career life-balance initiative.* Retrieved from www.nsf.gov/Career-life-balance

Nosek, B. A., Smyth, F. L., Sriram, N., Lindner, N. M., Devos, T., Ayala, A., . . . & Greenwald, A. G. (2009). National differences in gender–science stereotypes predict national sex differences in science and math achievement. *Proceedings of the National Academy of Sciences, 106*(26), 10593–10597.

Pajares, M. F. (1992). Teacher beliefs and educational research: Cleaning up a messy construct. *Review of Educational Research, 62*(3), 307–332.

Purdie-Vaughns, V., Cohen, G., Garcia, J., Sumner, R., Cook, J., & Apfel, N. (2009). *Improving minority academic performance: How a values-affirmation intervention works.* The Teachers College Record.

Roehrig, G. H., Kruse, R. A., & Kern, A. (2007). Teacher and school characteristics and their influence on curriculum implementation. *Journal of Research in Science Teaching, 44*(7), 883–907.

Rokeach, M. (1968). *Beliefs, attitudes, and values: A theory of organization and change.* San Francisco, CA. Jossey-Bass.

Shaver, A., Cuevas, P., Lee, O. & Avalos, M. (2007). Teachers' perceptions of policy influences on science instructions with culturally and linguistically diverse elementary students. *Journal of Research in Science Teaching, 44*(5), 725–746.

Sherman, D. K., Hartson, K. A., Binning, K. R., Purdie-Vaughns, V., Garcia, J., Taborsky-Barba, S., . . . & Cohen, G. L. (2013). Deflecting the trajectory and changing the narrative: How self-affirmation affects academic performance and motivation under identity threat. *Journal of Personality and Social Psychology. Advance online publication.* doi: 10.1037/a0031495

Sherman, D. K., & Cohen, G. L. (2006). The psychology of self-defense: Self-affirmation theory. In M.

P. Zanna (Ed.), *Advances in Experimental Social Physiology* (Vol. 38, pp. 183–242). San Diego. CA: Academic Press.

Simpkins, S. D., Davis-Kean, P. E., & Eccles, J. S. (2006). Parental socialization and children's engagement in information technology activities. Math and science motivation: A longitudinal examination of the links between choices and beliefs. *Developmental Psychology, 42*(1), 70–83.

Southerland, S. A., Gess-Newsome, J., & Johnston, A. (2003). Portraying science in the classroom: The manifestation of scientists' beliefs in classroom practice. *Journal of Research in Science Teaching, 40*(7), 669–691.

Spillane, J. P. (2012). The more things change the more they stay the same? *Education and Urban Society, 44*(2), 123–127.

Sweeder, R. D., & Strong, P. E. (May-June, 2012). Impact of a sophomore seminar on the desire of STEM majors to pursue a science career. *Journal of STEM Teacher Education, 13*(3), 52–61.

The President's Council of Advisors on Science and Technology. (2010). *Report to the President. Prepare and inspire: K-12 education in science, technology, engineering, and mathematics (STEM) for America's Future.* Retrieved from Washington, DC. Retrieved from www.whitehouse.gov/ostp/pcast

The President's Council of Advisors on Science and Technology (PCAST) (2012). *Report to the President. Prepare and inspire: K-12 education in science, technology, engineering, and math (STEM) for America's Future.* Washington, DC. Retrieved from: www.whitehouse.gov/ostp/pcast

Wallace, C. S., & Kang, N. (2004). An investigation of experienced secondary science teachers' beliefs about inquiry? An examination of competing beliefs sets. *Journal of Research in Science Teaching, 41*(9), 936–960. doi:10.1002/tea.20032

Wang, M., Eccles, J., & Kenny, S. (2013). Not lack of ability but more choice: Individual and gender differences in choice of careers in science, technology, engineering, and mathematics. *Psychological Science, 24*(5), 770–775. doi: 10.1177/0956797612458937

Washinawatok, K. (1993). Teaching Cultural Values and Building Self-Esteem. Wigfield, A. & Eccles, J. S. (2000). Expectancy-value theory of achievement motivation. *Contemporary Educational Psychology, 25*, 68–81.

Yerrick, R., Parke, H., & Nugent, J. (1997). Struggling to promote deeply rooted change: The filtering effectof teachers' beliefs on understanding transformational views of teaching science. *Science Education, 81*(2), 131–159.

AFFILIATION

Celestine H. Pea
Division of Research on Learning
National Science Foundation
Arlington, VA

JULIE A. LUFT & SISSY S. WONG

CONNECTING TEACHER BELIEFS RESEARCH AND POLICY: AN OVERVIEW AND POTENTIAL APPROACHES

CONNECTING TEACHER BELIEFS RESEARCH AND POLICY:
AN OVERVIEW AND POTENTIAL APPROACHES

This chapter illustrates why the examination of teacher beliefs is important in light of forthcoming and accepted educational policies. We discuss the interface between beliefs research and policy, propose a guiding model that links policy and teacher beliefs research, and suggest different research approaches in the context of this model. Throughout the chapter, we also offer examples of research that connects teacher beliefs and policies. We hope to initiate and advance the dialogue among researchers about this area of study, and to make a contribution to policies and beliefs research.

In order to begin a discussion about the connection of teacher beliefs research and policy, it is important to recognize the shifting nature of policies in teacher education. Around the world, heightened interest in the education and performance of teachers has resulted in new policies to guide teacher preparation and teacher professional development. In the United Kingdom (U.K.), for instance, standards have been developed that call for content knowledge, an understanding of student learning, a knowledge of assessment, planning and teaching, and professionalism (U.K. Department of Education, 2012). Each area of concentration contains a list of the specific competencies needed in order to meet the standards. The United States (U.S.) and Australia have adopted similar standards, with expanded descriptions that will be used to monitor teacher development (Australian Institute for Teaching and School Leadership, 2012; Council of Chief State School Officers, 2011). These descriptions offer a professional trajectory of learning, made up of distinct levels within each standard.

The international standards fall into the areas of content knowledge, an understanding of student learning, the knowledge of assessment, planning and teaching, and professionalism. They clearly suggest that teacher practices will impact student learning. In order to ensure that these standards are met, students and teachers may be evaluated on their knowledge and performance. In the U.S., for example, student assessments are prevalent in mathematics, English and science—a result of the No Child Left Behind Act of 2001 (NCLB). Similarly, teachers are required to pass state content and pedagogical knowledge assessments in order to receive their

R. Evans et al., (Eds.), The Role of Science Teachers' Beliefs in International Classrooms, 135–147.

teaching certificate. A movement towards assessing teacher performance in the U.S. has inspired a lively debate about the promises and pitfalls of such a system (see Darling-Hammond, Amerin-Beardsley, Haertel, & Rothstein, 2011).

While the research that connects standards and teacher practices is important, it overlooks the condition that guides the practices of teachers – their beliefs. We suggest that looking at the beliefs of teachers is important, as beliefs influence actions (Czerniak & Lumpe, 1996; Nespor, 1987; Pajares, 1992; Richardson, 1996). Even with support from researchers interested in understanding teacher beliefs, few have pursued this line of work to the degree that is needed, and rarely do researchers explicitly examine the beliefs of teachers in relation to current policies. With the global emphasis on teacher standards, in addition to research that examines the content knowledge and practices of science teachers, we will need further investigation into how teachers attain content-based standards in light of their beliefs.

THE INTERFACE OF POLICY AND BELIEFS

The connection of policy to research in science teacher education is relatively new. In 2001, White noted an absence of studies on policy research after reviewing decades of science education research studies. Fensham (2009) reviewed two years of the *Journal of Research in Science Teaching* and found only one article and one guest editorial that addressed policy work. Additionally, he analyzed two science education research handbooks and was surprised by the absence of any work focused on policy (Fensham, 2009). In fact, Fensham noted, "In the Fraser and Tobin handbook, the authors of the three papers on curriculum change and reform remarkably managed to avoid making any reference to the word 'policy' (p. 1077)." In this same article, Fensham (2009) called for connection of research, policy, and practice, as well as consideration of the many factors that influence the relationship between these three entities. Critical factors include the roles of stakeholders who are internal and external to school settings, as well as the impact of the values and authority of all those involved when linking policy and practice.

More recently, Luft and Hewson (in press) highlighted the presence of policy in the field of professional development program research. The connection between educational policies and various teacher factors occurred in studies that stated the standards (considered to be the policy area) as a goal of teacher learning, and then measured the changes experienced by teachers. Standards and national or regional issues were used to frame the problem, while teacher knowledge or teacher practice were common measures that indicated attainment of the standards. Unlike Fensham (2009), Luft and Hewson (in press) found studies that linked policy and teacher knowledge. However, they did not report on any studies from 2003-2012 that explicitly connected policies to teacher beliefs.

There is certainly a need for research that connects policy and teachers' beliefs, as beliefs are mediators of practice (Cimbricz, 2002; Davis, 2003; Mansour, 2008).

While the connection between policies and teachers' beliefs is assumed, it is not well understood. As a consequence, those who work with teachers to promote teacher change (which is often the outcome of such research) are hampered in supporting the design and implementation of teacher education programs that can impact teacher beliefs, and ultimately teacher practice. In order to understand the complicated interface between policy and teacher beliefs research, it is necessary to characterize the orientation of the policy, and the beliefs of teachers that potentially relate to the policy. Existing research in the area of policy and beliefs can clarify this necessary but complex connection. Figure 1 will help guide this discussion.

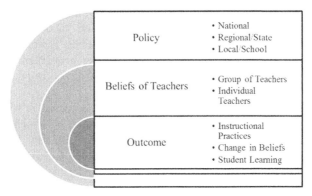

Figure 1. A model illustrating the connection of policy, teacher beliefs, and measured outcomes.

Figure 1 illustrates the nested nature of policy, beliefs, and the anticipated outcome. At the outermost level is policy. The policy can be a national document, a regional or state requirement, or a school or local rule. Specific policies can mandate, for example, the funding of professional development programs for teachers, the structure of an initial certification program for science teachers, or the number of professional development hours that a teacher must complete in a year. These policies can be stated by groups of people who are remotely familiar with education, or groups that are invested in the educational environment that is under study.

The intermediate level consists of science teacher beliefs. Ultimately, the size of the sample pool indicates the relationship of the policy to the beliefs of the teachers. Beliefs can be collected from a large group of teachers and in direct response to a specific policy. The beliefs held by the teachers are guided by their prior experiences in education, their understanding of students, or experiences outside of education. Studies collecting the beliefs of a large group of teachers tend to utilize standardized assessments, and to propose and test a theory about the connection of beliefs and policy. Beliefs discussed in a study may also come from an individual teacher or a small group of teachers. Studies that focus on a small number of teachers often try to understand the beliefs held by the teacher or teachers as they pertain to overarching policy. These studies suggest connections that can occur between policy and teacher beliefs. Again, the difference in sample sizes indicates a researcher's desire to understand the different direction of the relationship between policies and beliefs.

The interior level of Figure 1 consists of the outcomes of policy research, which takes into account teacher beliefs. These outcomes represent the effect of the stated policies. They can be reported as changes in instruction or beliefs of a teacher or a group of teachers, or changes in student learning. At times, the results of the teachers can be emphasized with student results addressed secondarily, or the results of students can be emphasized with the teacher results addressed secondarily. In either instance, when teacher or student outcomes are reported, they are directly connected to the policy, with some explanation about the nature of the connection.

In research that focuses on policy and teacher beliefs, the relationship between teacher beliefs and policy can take on different orientations. The orientations can be consistent teacher beliefs between teachers and policies, incongruent teacher beliefs and policies, and mixed teacher beliefs and policies. The following examples will clarify these different orientations.

In very few instances, the orientation of the policy is consistent with the beliefs of teachers. This type of research often reports how teacher beliefs changed to align with the stated policy. In a unique study in this area, Pilitsis and Duncan (2012) followed 17 preservice teachers as they engaged in their secondary science methods course. In this study, the authors were interested in understanding how preservice teachers re-oriented their beliefs in response to U.S. policy reform documents. They found that as the preservice teachers experienced different types of instructional approaches aligned with reform-oriented instruction, they developed more reform-oriented beliefs. They concluded that teachers' beliefs could be modified to comply with national reforms, but there was still more to learn about the process of modifying or changing teacher beliefs.

Another orientation in this research area is the incongruence between policy and teacher beliefs. When policy and teacher beliefs are incongruent, the policy is not enacted by teachers as envisioned. Tan (2011) illustrates this in a study of policy change in Malaysia. In 2003, Malaysia changed the national language of instruction to English, with the goal of increasing English proficiency and students' mathematical and science learning. Suddenly, science and mathematics teachers had to function as English, as well as content instructors. Tan (2011) found that when working on language during instruction, the science and mathematics teachers focused on defining concepts instead of building the students' capacity in language. This instruction limited the potential for language work in the science and mathematics courses. Tan (2011) also found that science and mathematics teachers believed they should focus on content of their subject areas, while language teachers believed they should focus on teaching the English language. This study revealed that the teachers' distinct beliefs about their roles in the classroom created challenges for adhering to the new national policy about learning content and language.

Another example of incongruence between policy and teacher beliefs is illustrated in a study by Milner, Sondergeld, Demir, Johnson, and Czerniak (2012). In this U.S. study, there is evidence of the difficult transition between the intended policy and teacher, and the teacher and student experiences. The goal of this study was to determine whether the emphasis on teaching science had changed in the elementary setting since the onset of NCLB. The NCLB policy envisioned a robust education for students in all academic areas, but only assessed students in the area of literacy and mathematics. Milner et al. (2012) collected qualitative data from 44 elementary teachers, and quantitative data from over 140 elementary teachers who were participating in a professional development program on teaching science. The collected data were analyzed and revealed that teachers' beliefs about teaching science remained unchanged despite NCLB, and that the teachers reported enacting less science since NCLB. In this case, the NCLB policy was instituted, but elementary teachers held beliefs that emphasized literacy and mathematics even in the midst of a science-focused professional development program. Ultimately, the teachers' beliefs were not receptive towards the NCLB policy in the non-tested area of science, and despite the policy's intent, students did not get additional science experiences.

Another orientation in this research area is the examination of congruent and incongruent teacher beliefs in the midst of the advocated policy. This is a 'mixed' orientation. Not surprisingly, there is more accumulated research in this area. An example of the varied beliefs that teachers can hold in the midst of a national policy can be found in Czerniak and Lumpe (1996). In this study, they investigated U.S. science teachers' beliefs about the *National Science Education Standards* (NSES) (National Research Council (NRC), 1996), and the implementation of strategies aligned with this reform document. The teachers in the study believed reform was needed, and most implemented strategies aligned with the NSES (NRC, 1996), such as cooperative learning. Even though there was support for the document, 81% of participants did not believe in the reform's central notion of constructivism, nor did they implement the constructivist strategies advocated in the document. In Czerniak and Lumpe's (1996) study, they found teachers' beliefs about reform to be the greatest indicator for implementing reform-based strategies. If a teacher did not believe in the necessity of changing the way science was taught and assessed, changes were not likely to occur in the their classroom instruction.

Davis (2003) also illustrates the mixed nature of teacher beliefs in the midst of an emphasized policy. The study examined how middle school science teacher beliefs influenced whether they implemented reform-based curriculum, which aligned with U.S. documents guiding science education. Davis (2003) found that not all the teachers' belief systems aligned with the reform-based curriculum materials. Teachers with more teacher-directed beliefs were not convinced the new curriculum materials would be an improvement when compared to the already established curriculum. These teachers experienced challenges in incorporating the new curriculum into their instruction. On the other hand, teachers with student-centered beliefs systems possessed greater knowledge of the concepts and strategies in the new curriculum and were more apt to integrate the material into their existing instruction. Ultimately, the beliefs of the teachers guided their use of the new curriculum, which resulted in their classroom practice aligning with advocated national reforms.

In summary, there is a connection between policy, teacher beliefs, and expected outcomes. Research that bridges these three areas is important, but it reveals that teacher beliefs can be congruent with policy, incongruent with policy, or that teachers can have beliefs that are both congruent and incongruent with policy. Research often reveals that teachers' beliefs are incongruent with advocated policies, or that teachers hold mixed beliefs about the advocated policies. The beliefs held by teachers ultimately have an influence on their instruction.

RESEARCH APPROACHES THAT HAVE POLICY IMPLICATIONS

As additional studies are conducted in the area of teacher beliefs, four research approaches will be especially useful to policy makers or those who guide their work: the synthesis of studies on beliefs, the use of valid instruments or measures, longitudinal studies of populations, and studies that explore the connection of beliefs and practice. Synthesizing the research on teacher beliefs is a common method to determine salient findings pertaining to teacher beliefs and policy. This approach requires the review of articles in order to make general and specific conclusions. Another approach involves the use of a common instrument to measure teacher beliefs. This approach allows for the consolidation of several studies in order to make a compelling case about the impact of, for example, an event or instructional approach. In this book, the chapter about the use of instruments is relevant to the present section. Another important approach pertains to longitudinal studies. These studies highlight trends in beliefs over time, in the midst of specific policies. Finally, studies can explore the connection between teacher beliefs and practices. While this connection is often assumed, more research in this area is certainly needed. The following sections will illustrate these different areas.

Synthesis Studies

When making decisions, policy makers are often interested in the accumulation of results. Results can be derived from a synthesis of several studies, or they can be the collective results of an instrument that has been used over time. This second point will be discussed in the next section. In terms of the first point, when enough data are collected it is possible to make decisions that guide certain policies. Data pertaining to the beliefs of teachers help us to avoid crafting policies that are incompatible with teacher beliefs, or that result in unexpected outcomes (Eisenhart, Cuthbert, Shrum & Harding, 1988).

Synthesis studies are often reviews of a collection of studies, although they can also be a statistical analysis of a collection of studies. Kang, Sandretto and Heath's (2002) synthesis study on beliefs and practices of tertiary instructors is an example of this approach. In this review, they examined 50 studies on the beliefs and practices of those in higher education. Among other findings, they concluded that there were inconsistencies between the beliefs and practices of tertiary instructors, and that conclusions not grounded in research had been drawn about the beliefs and practices of instructors. Their review of research demonstrated the need for professional development opportunities for academics in higher education in order to build their beliefs and practices in ways that align with high quality learning experiences.

For policy makers, Kang, Sandretto and Heath's (2002) study questions how faculty teach and how they are supported to teach. This study adds to numerous investigations of the questionable instructional conditions in higher education. While policy makers have not mandated professional development or certification for those in higher education, there is a movement in the U. S. to improve the educational experience of undergraduate science students. Several documents have been published in just the last 10 years that emphasize the need to improve the instruction of faculty, and learning of their students. The Vision and Change Report (American Association for the Advancement of Science, 2009), for example, calls for the improvement of biology instruction at all levels in the higher education system. Ultimately, this type of study could be used to support changes in higher education, which would involve both the crafting and funding of policies.

Instrument Use

Another way in which the accumulation of data can guide policy pertains to data collection. In science, self-efficacy is often monitored through the Science Teaching Efficacy Belief Instrument (STEBI) (Enochs & Riggs, 1990). The STEBI is a 23-item instrument that uses a rating scale from strongly agree to strongly disagree, and was developed for elementary teachers. Monitoring the change of science teachers' self-efficacy has been a long-standing area of interest among those who work with science teachers. The assumption associated with this instrument is that without good self-efficacy, important instructional practices will not be adopted by a science teacher.

Countless studies have been conducted using the STEBI, and they reveal the impact of self-efficacy on instruction. The most common type of study examines the effect of a professional development program on a teacher's self-efficacy. Palmer (2011), for example, studied the self-efficacy of 12 Australian elementary teachers who participated in a professional development program. Data were collected through interviews and the STEBI prior to, during, after, and two years after the program. The data were analyzed in order to understand the improvement of teachers in terms of their self-efficacy. The author concluded that the teachers' self-efficacy improved as they engaged in a professional development program that targeted their perceived abilities to teach science. In addition, self-efficacy improved as teachers were provided with feedback about their teaching.

In another study using the STEBI, Lakshmanan, Heath, Perlmutter and Elder (2011) explored how professional learning communities supported teacher self-efficacy. In addition to the STEBI, data were also collected through observations of practice. In this study, the self-efficacy of the elementary and middle school U.S. teachers did not significantly increase by the end of the program. However, teachers with higher self-efficacy tended to use practices advocated by the professional development program providers, while teachers with lower self-efficacy struggled to adopt the practices advocated in professional development program.

For policy makers, these studies could suggest professional development program formats that guide teacher beliefs, which would in turn have an effect on student learning. The studies also confirm the ongoing need to support the professional development of teachers, since without professional development the beliefs of teachers cannot be modified to support various goals of policy documents. In addition, these data provide information about the beliefs of teachers, which policy makers can monitor as they decide on policies or reform directions that are attainable by teachers. As in the previous area, decisions made about supporting teacher professional learning will have fiscal implications for policy makers.

Longitudinal Research

Longitudinal research is important in teacher belief and policy research. This type of research demonstrates how beliefs can change over time in the presence of policy-related initiatives (which may be at different educational levels). An example of longitudinal research that focuses on teacher beliefs can be found in the work of Luft and her colleagues. They followed close to 100 beginning U.S. secondary science teachers over five years in order to understand how the beliefs of teachers changed as they learned about standards-based instruction. During the first two years, however, the new teachers experienced different types of induction programs. Two programs emphasized reform-based science, while two programs did not have this emphasis. The science instruction advocated in the science induction program aligned with the National Science Education Standards (NRC, 1996).

The participating teachers were interviewed using the Teacher Belief Interview (TBI) (Luft & Roehrig, 2007) prior to their first year of teaching, and then each year afterwards for five years. The responses of the teachers were quantified following Miles and Huberman (1994). Traditional and instructive responses represented more traditional or teacher-centered beliefs, and were scored one or two respectively. Responsive and reform-based responses represented beliefs aligned with the goal of the current science education reforms and student-centered learning, and were scored with four or five respectively. Transitional responses, scored with a three, demonstrated an affective response toward students, but did not clearly affirm students' roles in the classroom as co-constructors of knowledge. The responses for each participant to the questions on the TBI were summed and used in the analysis. A total score of 35 indicates student-centered beliefs, while a score of 7 indicates teacher-centered beliefs. Scores in the middle represent beliefs that are moving between teacher and student-centered orientations.

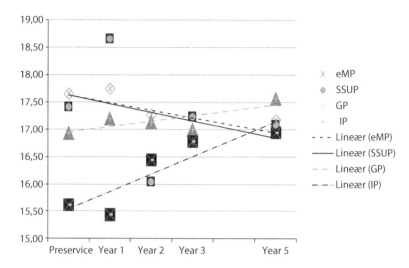

Figure 2. TBI scores and best fit lines (linear) of beginning teachers in the different induction programs over five years. eMP = electronic Mentoring Program, SSUP = Science Specific University Program, GP = General Program, IP = Intern Program.

Figure 2 shows how the beginning teacher beliefs changed over time. It indicates that the beginning secondary science teachers held different beliefs about inquiry instruction prior to starting their careers. Over time, their beliefs became similar and they collectively held transitional beliefs about inquiry.

To better understand the belief changes of the beginning science teachers over this five year period, qualitative data were also collected. The data were analyzed thematically and revealed that induction programs had an impact in the early years on the beliefs of the new teachers. Over time, however, the new teachers were subsumed into the school's belief system about the use of inquiry instruction.

From a beliefs and policy perspective, this study demonstrates the fluctuating nature of teacher beliefs and the need to provide ongoing professional development opportunities to new teachers in order to ensure that they are supported to enact reform-based instruction. Longitudinal research can reinforce the need for professional learning at different career stages, and it can suggest different ways in which teachers develop. For policy makers, this study and other longitudinal studies, suggest that there is a need to consider how to support science teachers throughout their careers. Specifically, adequate guidelines should be established and adequate funds should be allocated in order to provide teachers with various educational opportunities to support and strengthen their instruction and beliefs orientations.

Teacher Beliefs and Practices

Another important research area to consider is the relationship between teachers' beliefs and practices. In 1987, Nespor stated that beliefs "play a major role in defining teaching tasks and organizing the knowledge and information relevant to those tasks" (p. 324). Since then, research has shown that a teacher's beliefs influence classroom practice. For example, Brickhouse's (1990) study of three U.S. science teachers reported that the beliefs of the teachers about the nature of science guided their design and implementation of science lessons. Similarly, Cronin-Jones's (1991) work with two U.S. middle-grade science teachers found that both teachers' beliefs about student learning and their own roles in the classroom influenced the ways they adapted and implemented curriculum.

When studying the connection of beliefs and practice, it is important to observe the practices of teachers. Several researchers in this area do observe teacher practices while monitoring their beliefs. Luft, Roehrig and Patterson (2003) followed 18 beginning secondary science teachers into their first year of teaching. They observed and interviewed the new teachers in order to understand how their beliefs and practices were impacted by their participation in an induction program. One group of teachers participated in an induction program focused on science, and another group participated in an induction program developed by a school district. The last group of teachers did not participate in an induction program. In a multiple methods study, Luft et al. (2003) found that the new teachers in science-focused induction program implemented more inquiry oriented investigations and had more student-centered beliefs than did their counterparts in the other programs. They concluded that science-focused induction programs were important in supporting beginning science teachers.

These studies, and others in this area, are important because they link beliefs and practices. However, studies that rely on the self-report of teacher instruction do not establish a firm link between beliefs and practice. If future studies in this area are going to impact policy, they will have to include researcher observations of teachers. Unfortunately, this type of work falls into a cycle: Funding is needed in order to ensure the research happens, but compelling research needs to be present in order to secure funding. For policy makers, the connection of beliefs and practice is apparent, but more data are needed.

NEXT STEPS: BRINGING IT ALL TOGETHER

Teachers are ultimately responsible for implementing educational policies in the classroom, therefore they should be involved in the policy making process. Some academics advocate for teachers to be included as equal collaborators in decision making (Eisenhart et al., 1988). They reason that including teachers in policy formation and implementation is critical because it may influence their beliefs about the policy, and the teachers may also gain a better understanding of the purpose of the policy. Most academics, however, use research that examines the beliefs or practices of teachers to speak to policy makers. In doing this, teachers are not presenting their views to policy makers. Instead, academics are presenting the views of teachers, which may or may not be consistent with the actual views of teachers. In the area of beliefs, several parameters need to be considered when bridging policies and teacher beliefs. As discussed in this chapter, they include understanding the connection of a policy to teacher beliefs, and the type of research approaches that will be promising in the area of policy.

No matter how teachers are included in future policy work, research pertaining to teacher beliefs must take a more strategic approach in order to impact policies that pertain to science teachers. While we have suggested ways in which to enhance research in the area of beliefs and policy, we should also point out that simply understanding how a study can relate to policy is crucial. Educational researchers should be able to show how their research directly relates to a policy. Furthermore, this relationship needs to be foregrounded, either at the beginning of the study, or in a special section that links the policy to the problem that is under investigation. In doing so, researchers provide policy makers with one more signpost to guide their decision-making process.

AUTHORS' NOTE

The authors would like to recognize the following faculty and research assistants who helped with some of the data reported in this chapter: Krista Adams, EunJin Bang, Jonah Firestone, Elizabeth Lewis, Jennifer Neakrase, Irasema Ortega, Charles Weeks and Gillian Roehrig. This chapter and some of the data reported upon were made possible by National Science Foundation grants 0550847, 0918697, 0732600, and 0632368. The findings, conclusions, and opinions herein represent the views of the authors and do not necessarily represent the view of personnel affiliated with the National Science Foundation.

REFERENCES

American Association for the Advancement in Science (2009). *Vision and change in undergraduate biology education: A call to action.* Washington DC: Author.

Australian Institute for Teaching and School Leadership (2012). *Australian professional standards for teachers.* Education Services Australia. Retrieved from http://www.teacherstandards.aitsl.edu.au/Standards/AllStandards

Brickhouse, N. (1990). Teachers' beliefs about the nature of science and their relationship to classroom practice. *Journal of Teacher Education, 41*(3), 53–62.

Cimbricz, S. (2002). State-mandated testing and teachers' beliefs and practice. *Education Policy Analysis Archives, 10*(2), 1–21.

Council of Chief State School Officers (2011). *Interstate teacher assessment and support consortium (InTASC) model core teaching standards: A resource for state dialogue.* Washington, DC: Author.

Cronin-Jones, L. L. (1991). Science teacher beliefs and their influence on curriculum implementation: Two case studies. *Journal of Research in Science Teaching, 28*(3), 235–250.

Czerniak, C., & Lumpe, A. (1996). Relationship between teacher beliefs and science education reform. *Journal of Science Teacher Education, 7*(4), 247–266.

Darling-Hammond, L., Amerin-Beardsley, A., Haertel, E., & Rothstein, J. (2011). *Getting teacher evaluation right: A challenge for policy makers.* American Education Research Association & National Academy of Education, Research briefing. Retrieved September 14, 2011, from http://www. aera.net/Portals/38/docs/New%20Logo%20Research%20on%20Teacher%20Evaluation%20AERA-NAE%20Briefing.pdf

Davis, K. S. (2003). Change is hard: What science teachers are telling us about reform and teacher learning of innovative practices. *Science Education, 87*(3), 3–30.

Eisenhart, M., Shrum, J., Harding, J., & Cuthbert, A. (1988). Teacher beliefs: Definitions, findings and directions. *Educational Policy, 2,* 51–70.

Enochs, L., & Riggs, I. (1990). Further development of an elementary science teaching efficacy belief instrument: A preservice elementary scale. *School Science and Mathematics, 90,* 694–706.

Fensham, P. J. (2009). The link between policy and practice in science education: The role of research. *Science Education, 93*(6), 1076–1095.

Kang, K., Sandretto, S., & Heath, C. (2002). Telling half the story: A critical review of research on the teaching beliefs and practices of university academics. *Review of Educational Research, 72(2),* 177–228.

Lakshmanan, A., Heath, B. P., Perlmutter, A., & Elder, M. (2011). The impact of science content and professional learning communities on science teaching efficacy and standards-based instruction. *Journal of Research in Science Teaching, 48,* 534–551.

Luft, J. A., Firestone, J. B., Wong, S. S., Ortega, I., Adams, K., & Bang, E. J. (2011). Beginning secondary science teacher induction: A two-year mixed methods study. *Journal of Research in Science Teaching, 49*(10), 1199–1224.

Luft, J. A., & Hewson, P. W. (in press). Research on teacher professional development programs in science. In S. K. Abell & N. Lederman (Eds.), *Handbook of Research in Science Education.*

Luft, J. A., & Roehrig, G. H. (2007). Capturing science teacher's epistemological beliefs: The development of the teacher beliefs interview, *Electronic Journal of Science Education, 11*(2). Retrieved from http:// ejse.southwestern.edu/volumes/v11n2/articles/art03_luft.pdf

Luft, J. A., Roehrig, G., & Patterson, N. C. (2003). Contrasting landscapes: A comparison of the impact of different induction programs on beginning secondary science teachers' practices, beliefs, and experiences. *Journal of Research in Science Teaching, 40*(1), 77–97.

Mansour, N. (2008). Science teachers' beliefs and practices: Issues, implications and research agenda. *International Journal of Environmental & Science Education, 4*(1), 25–48.

Miles, M. B., & Huberman, A. M. (1994). *Qualitative data analysis.* Thousand Oaks, CA: Sage.

Milner, A. R., Sondergeld, T. A., Demir, A., Johnson, C. C., & Czerniak, C. M. (2012). Elementary teachers' beliefs about teaching science and classroom practice: An examination of pre/post NCLB testing in science. *Journal of Science Teacher Education, 23*(2), 111–132. Retrieved from http://login. ezproxy1.lib.asu.edu/login?url=http://search.proquest.com/docview/1011395880?accountid=4485

National Research Council [NRC]. (1996). *National science education standards.* Washington DC: National Academy Press.

Nespor, J. (1987). The role of beliefs in the practice of teaching. *Journal of Curriculum Studies, 19*(4), 317–328.

No Child Left Behind Act of 2001, 20 U.S.C. § 6319 [NCLB] (2008).

Pajares, M. F. (1992). Teachers' beliefs and educational research: Cleaning up a messy construct. *Review of Educational Research, 62*(3), 307–332.

Palmer, D. (2011). Sources of efficacy information in an inservice program for elementary teachers. *Science Education, 95,* 577–600.

Pilitsis, V., & Duncan, R. G. (2012). Changes in belief orientations of preservice teachers and their relation to inquiry activities. *Journal of Science Teacher Education, 23*(8), 909–936.

Richardson, V. (1996). The role of attitudes and beliefs in learning to teach. In J. Sikula (Ed), *The handbook of research in teacher education* (2nd ed.), (pp. 102–119). New York, NY: Macmillan.

Tan, M. (2011). Mathematics and science teachers' beliefs and practices regarding the teaching of language in content learning. *Language Teaching Research, 15*(3), 325–342.

U.K. Department of Education. (2012). *Teachers' standards, May 2012.* Crown copyright. Retrieved from http://webarchive.nationalarchives.gov.uk/20130401151715/ https://www.education.gov.uk/publications/standard/publicationDetail/Page1/DFE-00066-2011

White, R. T. (2001). The revolution in research on science teaching. In V. Richardson (Ed.), *Handbook of research on teaching* (4th ed., chap. 25, pp. 875–898). Washington, DC: American Educational Research Association.

AFFILIATIONS

Julie A. Luft
Department of Mathematics and Science Education
College of Education
University of Georgia

Sissy S. Wong
Department of Curriculum and Instruction
College of Education
University of Houston

SECTION FOUR

BELIEFS RESEARCH AND TEACHERS

SARAH ELIZABETH BARRETT

BECOMING AN ACTIVIST SCIENCE TEACHER: A LONGITUDINAL CASE STUDY OF AN INDUCTION INTERVENTION

INTRODUCTION

This longitudinal case study is part of a larger multi-case study examining the relationship between new science teachers' espoused and enacted beliefs over the course of time. Although teachers' espoused beliefs are known to significantly affect their teaching practice (enacted beliefs) (Brock & Boyd, 2011), very little is known about the path from one to the other (Bryan, 2003; Luft, 2007). This chapter seeks to address that gap by focusing on one high school teacher's beliefs with regard to teaching for social justice by teaching about and through socioscientific issues.

Socioscientific issues are science-related issues that have ethical overtones and require moral reasoning to address (Zeidler & Sadler, 2008). They cover a wide range of topics, including but not limited to bioethics, research funding priorities, environmental issues, local and global socioeconomic issues, and geopolitical issues. These topics become social justice issues when they relate specifically to redistribution of wealth and knowledge that arise through science practice and recognition of the contributions and aspirations of people who are not from the dominant culture (Atwey, 2011). An activist science teacher who is interested in social justice would be "able to concretely and knowledgeably bring knowing to the problems at hand" (Roth, 2010) by using socioscientific issues to encourage students to consider and engage in this science-related redistribution and recognition while initiating change within the school (Reeves, 2007; Sachs, 2003).

Hodson (2003) has described four levels of increasing sophistication for the inclusion of socioscientific issues in science education:

Level 1: Appreciating the societal impact of scientific and technological change, and recognizing that science and technology are, to some extent, culturally determined.

Level 2: Recognizing that decisions about scientific and technological development are taken in pursuit of particular interests, and that benefits accruing to some may be at the expense of others. Recognizing that scientific and technological development are inextricably linked with the distribution of wealth and power.

R. Evans et al., (Eds.), The Role of Science Teachers' Beliefs in International Classrooms, 151–164.
© *2014 Sense Publishers. All rights reserved.*

Level 3: Developing one's own views and establishing one's own underlying value positions.

Level 4: Preparing for and taking action. (p. 655)

These levels can refer to three different aspects of teaching-learning: (1) the teacher's purpose, (2) the teacher's curricular choices, and (3) the student's learning. For example, the teacher's purpose may be at Level 4, curricular choices may be at Level 3, while the student's learning is at Level 2. This complexity in the same instance of teaching-learning suggests that teachers' espoused and enacted beliefs about how to teach a particular topic are highly contingent.

WEB OF BELIEFS

Luft (2009) defines beliefs as "propositions held to be true by the individual; they can be non-evidential and based on personal judgment and experience, unlike knowledge that is evidential and requires community or group consensus" (p. 2358). However, while this definition adequately describes espoused beliefs, it does not include action. Following Kane, Sandretto, and Heath (2002), I distinguish between espoused and enacted beliefs, the former being those beliefs that people express verbally and the latter being actions based on beliefs. For new teachers, this is an especially important distinction because new teachers may not necessarily have the skills to enact the beliefs they espouse.

There is a tendency in research on teacher beliefs to use enacted beliefs to judge the strength with which espoused beliefs are held (see, for example, Barrett & Nieswandt, 2010; Theriot & Tice, 2009), but this approach may be an over-simplification that masks the struggles teachers have with reconciling their espoused and enacted beliefs. For any given teacher, there may not be a straight line between espoused and enacted beliefs, partly due to the teacher's position within an institution that has its own trajectory (Barrett, Ford, & James, 2010). To understand the relationship new between teachers' changing (or unchanging) espoused and enacted beliefs, it is important to follow the actual trajectory over time (Fletcher & Luft, 2011).

A new teacher is unlikely to have experienced teaching through social justice–related socioscientific issues either as a student or during practice teaching. Research suggests that whether or not science teachers choose to teach differently than they have experienced probably depends on four factors: (1) their sense of the degree to which the culture of the science department supports their efforts (Friedman, Galligan, Albano, & O'Connor, 2009; Milner, Sondergeld, Demir, Johnson, & Czerniak, 2012), (2) their confidence in their pedagogical content knowledge (PCK) (Lumpe, Czerniak, Haney, & Beltyukova, 2012; Topcu, Sadler, & Yilmaz-Tuzun, 2010), (3) their conception of the nature of science (NOS) (Barrett & Nieswandt, 2010), and (4) their professional identity (Barrett & Nieswandt, 2010; Wenger, 1998). These four factors can be imagined as nodes in a web of personal and professional beliefs that guide teachers' curricular choices and, as such, can manifest as words (espoused beliefs) or actions (enacted beliefs) (Kane et al., 2002). (see Figure 1).

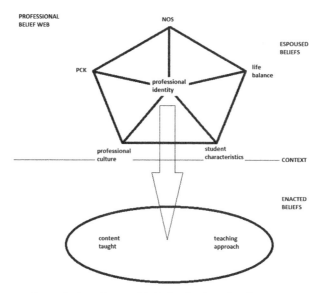

Figure 1. A professional belief web – based on literature.

Figure 1 shows a professional belief web. Espoused beliefs form a web with professional identity at its centre. Professional culture and student characteristics, which relate to context, are located on the border between espoused and enacted beliefs. Enacted beliefs include both the content that the teacher teaches as well as his or her approach to teaching. The trajectory between espoused and enacted beliefs is represented by an arrow that is anchored by professional identity. While this arrow is depicted as straight, this is not meant to imply a static trajectory. Shifts in other nodes on the web shift the trajectory: It can curve or it can land differently within the domain of enacted belief. Nor should the web as a whole be viewed as static. Rather, it is constantly changing shape to accommodate shifts in priorities with respect to different espoused beliefs, pulling in one direction or another, stretching and contracting.

New teachers' beliefs about the professional cultures in which they work affect their curricular choices (Friedman et al., 2009). Professional culture is communicated to them both formally and informally. In cases where formal teacher induction is provided, the goal tends to be to help new teachers to become familiar with procedures and workplace culture (Luft, 2009). Indeed, discussions of equity or social justice are rarely part of the agenda in formal induction (Barrett, Solomon, Singer, Portelli, & Mujawamariya, 2009; Bianchini & Brenner, 2010). However, regardless of any formal professional development that new teachers might participate in, it is the informal induction that they receive from colleagues with whom they work that has the biggest influence on their beliefs (Luft, Bang, & Roehrig, 2007; Milner et al., 2012). Thus, the enacted beliefs of new teachers espousing a belief in the necessity of teaching social justice through socioscientific issues may be highly influenced – positively or negatively – by the professional culture in which they work.

The skills new teachers must learn to teach effectively are part of professional content knowledge (PCK). PCK has been described as "an experiential knowledge that is acquired as a teacher works with students in the classroom, and is an integrated set of knowledge, conceptions, beliefs, and values that teachers develop in the context of a teaching situation" (Luft, 2009, p. 2359). Formal and informal induction will influence the development of PCK, but time spent witnessing and practising relevant skills within the classroom will be more influential. Indeed, teachers' own past experiences as high school students are important in this context. Since, as students, they were unlikely to have studied social justice–related socioscientific issues as part of their own science education, they may now – as new teachers – lack vision of how it might look. Also, they may not have a clear idea of the skills they need to develop, rendering development of PCK in this area problematic.

Closely related to PCK are teachers' beliefs about the nature of science (NOS). Much has been written about the significance of teachers' understanding of NOS (see, for example, Donnelly & Argyle, 2011). A teacher who believes science is an acultural activity based on a singular linear algorithm of inquiry will teach differently than a teacher who believes science is socially constructed and practised using a variety of approaches. The former understanding of NOS appears to be the usual approach in K–12 (Donnelly & Argyle, 2011) (although whether most teachers believe this interpretation of NOS is accurate is up for debate). Such an approach to NOS has little room for social justice–related socioscientific issues, which aims for activism on the part of teachers and students. In Figure 1, I have connected espoused beliefs about NOS, PCK, professional culture, and life-balance through professional identity because teachers' beliefs about who they aspire to be provide meaning to the trajectory their espoused beliefs take to enacted ones (Barrett & Nieswandt, 2010).

According to Luehmann (2007):

- Identity is socially constituted, that is, one is recognized by self and others as a kind of person because of the interactions one has with others.
- Identity is constantly being formed and reformed, though the change process for one's core identities is long term and labor intensive.
- Identity is considered by most to be multifarious, that is, consisting of a number of interrelated ways one is recognized as a certain kind of person, participating in social communities.
- Identity is constituted in interpretations and narrations of experiences. (p. 827)

Lasky (2005) sees teacher professional identity as "how teachers define themselves to themselves and to others." Helms (1998) noted that beliefs are at the core of identity. This is why professional identity is in the centre of the belief web in Figure 1 – because all espoused and enacted beliefs are connected through it.

DESIGN

This case study is part of a larger longitudinal multi-case study of physics and chemistry teachers in their early years of teaching. I utilized a case study approach because my purpose is to provide a detailed illustration of a complex "contemporary phenomenon within a real life context" that cannot be separated from that context (Kelly-Jackson & Jackson, 2012; Merriam, 1998, p. 27; Yin, 2009).

Data collection began at the end of Richard's first year of teaching. I was a participant-observer because I provided resources and feedback to Richard and participated in classroom activities when appropriate (Glesne, 2011). During Project Years 1-3, email correspondence and course materials were collected from Richard and he was interviewed bi-annually. During Project Year 4, classroom visits were added. All interviews were audio recorded and fully transcribed.

Data were examined for emerging themes (Strauss & Corbin, 1990) which were collapsed into larger themes through a process of decontextualization and recontextualization (Tesch, 1990). These larger themes were then compared to the literature and verified by the participant. Since all beliefs are deeply personal, I also used a cooperative approach to the final analysis, asking Richard to examine Hodson's (2003) framework to determine if he felt that it outlined valid benchmarks for examining his espoused and enacted beliefs. Richard and I then used these levels to discuss his work with his students. Together, we determined the levels of sophistication of his espoused and enacted beliefs. Finally, a professional belief web specific to Richard was constructed and verified by him.

Validity was established through (1) extended engagement with the participant, (2) multiple data sources, (3) member-checking of both the interpretations and the means of analysis, (4) discussions with other researchers in the science education field, (5) maintaining an electronic log of data, and (6) preliminary analysis using NVivo computer software.

BELIEFS ABOUT STUDENT CHARACTERISTICS: AN ALTERNATIVE HIGH SCHOOL PROGRAM

Since graduating from the faculty of education, Richard has been working in a publicly funded alternative high school program for students with attendance problems. The students are drawn from 10 to 15 nearby high schools. The entire program has nine locations with one principal who is not on Richard's site. Richard's curriculum leader for science and math is not at this site, either. Instead, the day-to-day program is run by Richard, one other teacher and an education assistant.

Each of the 15 to 20 students in the program arrives with a history of disengagement from school and often many problems at home. Because of their tendency to skip class, they are supervised all day. The program is structured so that they can focus on science and/or math (taught by Richard) in the morning and on English, history, and/or geography in the afternoon. Students work on their individualized programs at their own pace, aiming to pass four courses each semester.

Richard's science classroom is equipped with a counter at the back of the room with a single sink. He has limited equipment. There are whiteboards at the front of the classroom and four computers at the side of the room. The atmosphere in the school is relaxed – Richard generally wears jeans and a T-shirt or sweatshirt and students call him by his first name.

At the beginning of this study, Richard's espoused belief was that students' greatest need was to be heard both in school and in their lives. However, his students seemed too overwhelmed to have anything to say:

If you ask them [students] what their plan is for their future, they don't have a plan. They don't think there's a future and that's what's been really difficult to try and get around this year: It's to try and help them understand that they can make a change, and that they have [the] power to do things. (Spring 2007)

This challenge continued into subsequent years as he struggled to motivate students who "had never felt or seen success" (Spring 2007). Yet Richard expressed admiration for their resilience, saying, "I've been kind of stupefied by how they can take on those challenges at home and still come to school and work." (Fall 2010)

Four main themes emerged over the course of the study: (1) a supportive professional environment, (2) the elusive big picture, (3) aspiring to be a teacher activist, and (4) managing expectations.

BELIEFS ABOUT PROFESSIONAL CULTURE: A SUPPORTIVE PROFESSIONAL ENVIRONMENT

Richard's activism was rooted in feminist pedagogy and a commitment to leveling the playing field for women and other marginalized minorities (Pinnick, 2008). From the beginning, he felt supported in his social justice goals because he, the other teacher and the education assistant regularly discussed how best to do so. The result of these discussions was a cumulative project that replaced final exams in all subjects except science and math. Developed in consultation with the students themselves, the two-week project was designed to empower the students by helping them to focus on problems and causes bigger than themselves. In this way, the program's approach to teaching seems to be in line with Atwey's (2011) social justice focus on recognition and redistribution. However, even though Richard's colleagues were supportive of his social justice goals, that support was not specific to science.

Richard's espoused beliefs with respect to science focused on recognition. These beliefs were also consistent with Hodson's (2003) Level 4 – preparation for and taking action. Richard said that he wanted to get his students to that level of sophistication as well, but struggled with enacting that belief. As the only science and math teacher on site, Richard did not have the benefit of a colleague knowledgeable in teaching science and math on site to help him in his early years – something that new teachers cite as the most helpful for professional development (Luft, 2009). Although Richard noted that he felt he had the support of his curriculum leader and the principal, he still felt isolated because they were not on-site. When the curriculum leader position opened up, he expressed some optimism about his situation:

I think that I just really want to get going on it and do it and that's why I think applying for this job [will help]. I'm excited about being able to have a group of people to help me because I think I was overwhelmed. (Fall 2010)

Yet, while Richard's (and his colleagues') espoused beliefs were consistent with social justice education through feminist pedagogy and with Level 4 in Hodson's (2003) scheme, his enacted beliefs were not. Thus, although his professional culture was supportive, other factors – to be described in subsequent sections - intervened.

BELIEFS ABOUT PCK: THE ELUSIVE BIG PICTURE

Throughout the study, Richard sought the "big picture," a comprehensive view of the curriculum that would allow him to recognize appropriate spaces within the curriculum for presenting social justice–related socioscientific issues. He expressed gratitude for being lucky enough to work within a program that allowed for so much flexibility in program planning, but he was challenged in his first years by getting himself oriented in his new position as a full-time teacher. As he put it: "I didn't know what was going on this year, so I had a [hard] time seeing the big picture." (Summer 2007).

Having a big picture of the curriculum is part of PCK. A teacher with a comprehensive understanding of the curriculum she teaches not only knows the content but also how each topic relates to other topics in past and future studies. They are able to prioritize teaching–learning goals and to recognize appropriate enrichment opportunities. Such a vision is difficult in the early stages of a science teacher's career and developing it requires support from senior colleagues with expertise in teaching science. As participant-observer, I often offered such support, but, in the early years, Richard did not generally have enough of a handle on the curriculum to know what specific help he needed. By his third year, however, he was beginning to feel more confident:

Before, if I had a socioscientific issue, I couldn't look through and say "I can use that expectation" or "I can use this for this expectation." [Before], I never really had any connectivity between the whole thing. I never really saw it [the curriculum] as a whole document, [but] I feel like I really can see that now after looking through it a number of times and teaching the expectations a number of times. (Spring 2009)

However, he still found it difficult to bring his new vision to fruition, given his responsibilities with a growing family at home and the immediacy of day-to-day responsibilities at the school:

You get so caught up in all of these things that are happening all the time... final projects and running around like a chicken with their head cut off and trying to make sure you get a unit done and that a student has the next lesson ready because they're waiting for it and that sort of stuff. You kind of forget what you're really wanting to do and the direction that you really want to go in. (Fall 2010)

This struggle to find the mental space to think about goals that went beyond the day-to-day was a constant challenge for Richard, even as he gained confidence in his purpose and his vision of the curriculum. This problem is not unique to new teachers, but Richard's experience of this struggle was filtered through his aspirations to be an activist science teacher.

BELIEFS ABOUT NOS AND PROFESSIONAL IDENTITY:
BECOMING AN ACTIVIST SCIENCE TEACHER

Throughout the study, Richard tried to incorporate social justice-related socioscientific issues into his teaching of science and math. He said:

I really want them to get the idea that they can be critical and that they can own knowledge and come up with new ideas and question those people that are coming up with decisions about the things that affect their lives. (Summer 2007)

Richard noted that his students' experiences could be used as a jumping-off point for discussions about recognizing the contributions of non-dominant groups to science:

I think most of them [the students] had been thought of as being left behind, forgotten in some way or another. So I think they could really relate well to the people who have no voice. (Spring 2008)

This focus on recognition, while consistent with Atwey (2011), is only on Level 1 – applying and recognizing issues – in Hodson's (2003) scheme. It does indicate, however, a concept of NOS that emphasizes the universality of science across cultures while recognizing the role ethnocentrism has played in the dissemination of scientific knowledge from one culture to another.

As time went on, Richard espoused a commitment to inching students toward Hodson's Level 2 – recognizing vested interests in science/technology decision-making – by encouraging them to think critically about the reasons they were learning what they were learning. Richard's approach to encouraging critical thinking in his students during his first few years of teaching was ad hoc, pointing out the socially and politically contingent nature of science content in the curriculum in casual discussions with students as they worked:

Like to say: "Whose science are we learning? Where is that coming from? Why is it that we're learning science that has been 'developed' by these rich white guys?" . . . . [This approach works] because they're all students that have been marginalized in one way or another, in their lives. Even the ones that have come from what seems like fairly stable homes . . . they've been marginalized in one way or another. (Winter 2009)

Here, Richard notes the ways that these students' personal experiences help them to understand what he is trying to teach them.

In year three, in addition to these casual conversations with his students, he also added content related to socioscientific issues to the curriculum. For example, he added such topics as chemical dumping on First Nations' reserves, society's oil dependency, scientists' responsibilities relating to the results of their research, and the hegemony of Western science. These topics were a formal part of the curriculum and not just casual topics of conversation. As students worked through the materials he had prepared, Richard would try to engage them at a personal level through informal conversation about the material.

By Project Year 4, Richard was making attempts to push his students to Hodson's (2003) Level 3 – developing one's own views and establishing one's own value positions – through assignments within particular science units. Yet, as he and I discussed his accomplishments while preparing this paper, Richard noted that he really did not feel his students got past Level 1.

In his fifth year, Richard became the curriculum leader for science and math, responsible for coordinating eight science and math teachers at eight different sites. I mentioned earlier that he was very encouraged by this new responsibility and had suggested that he and his colleagues could collectively implement a curriculum that would fulfill his social justice goals.

BELIEFS ABOUT LIFE BALANCE: MANAGING EXPECTATIONS

Midway through Year 2, I asked Richard to imagine what advice he would like to have given to himself when he first started teaching:

The advice I think I would've given myself was to know that I would be overwhelmed, and that being overwhelmed does not mean that I am never going to be able to go back to including the things that I like to include, like social justice topics, in my science. And [I also know] that I am not going to be able to get it done overnight, but piece by piece I am going to be able to include these things in my work. (Winter 2008)

As time wore on, his optimistic and confident attitude was replaced by frustration, as evidenced by the following reflection a year later:

Maybe one thing that is getting in the way of it is inertia – that idea of staying in the same situation and you've got all your courses made up. And it's like you're running at Mach 5 during the run of the day because there's no such thing as prep time with our students. From 9 in the morning until 2:30 in the afternoon

– and that includes lunch; we sit together and eat lunch together, supervising them all day. So it's easier when you're running around like a chicken with its head cut off to just go ahead and grab the work that you've already made and hand it off to a student that's just arrived with their parents that you have done an intake with. It's kind of a crazy life at times. (Winter 2009)

A year later, Richard again considered his advice to himself as the brand new teacher:

Don't try and do everything at once. It'll come online. You just have to develop little things at a time and start incorporating them within . . . and it doesn't all have to be through curriculum – just the way you carry yourself in the classroom and talk about current issues in the classroom. . . . I think that as a new teacher coming in, don't be too uptight about what's expected of you from the Ministry [of Education] and try and develop your own style of how you're going to introduce these socioscientfic issues into what you're doing every day. (Summer 2009)

By the end of Year 4, Richard still alternated between acceptance of his limitations and frustration with his perceived lack of progress. When asked how I might help, he responded:

I think it's already being done with regards to just being able to reflect. It's a really good time of year right now to sit there and reflect while it's fresh in my head about what's happened in the past year. I felt like I kept letting myself down, especially after I'd come to the interviews and I'd talk to you and be reminded of all these things that were my goals, and I'd start to get kind of upset with myself that I haven't done any of that stuff yet. (Spring 2010)

This last quote is curious because Richard had done quite a lot of what he had said he wanted to do. Approximately half of the science unit plans I examined contained assignments revolving around socioscientific issues. However, they did tend to be at Level 2 in Hodson's (2003) scheme, in spite of the fact that Richard's espoused beliefs remained at Level 4. Richard admitted that it was this discrepancy between his espoused and enacted beliefs that was bothering him.

When Richard took on the role of curriculum leader in the beginning of his fifth year of teaching, he expressed a great deal of optimism about accomplishing his goals because his colleagues seemed enthusiastic about introducing more social justice-related socioscientific issues into the curriculum. However, their group curriculum writing sessions always ended up focusing on day to day concerns such as students' emotional needs and challenging behaviour, instead. Richard expressed frustration about this and his own inability to do what he wanted to do because of his increased daily workload due to his new responsibilities.

Furthermore, in anticipation of the arrival of his second child, Richard was also tutoring students outside work so that he could save up for a house. I have described all of these jobs and responsibilities to emphasize the range of difficulties (personal as well as professional) that Richard encountered in translating his espoused beliefs – which did not change over the four years of the study – into enacted ones. Richard evaluated his accomplishments in the following way:

> I feel that my role as curriculum leader is going OK, but I feel like although I had the greatest of intentions and dreams of transforming some of the ways we deliver curriculum, it just didn't fly the way I wanted it to and certainly didn't move along at a pace I tried to get everyone to set. I guess it was my intentions and dreams and not [those] of my colleagues. I am not sure if the work will ever get done unless I do it all myself . . . . I think that my colleagues are interested, but not to the same extent as me. I think that the pressures of just keeping up with the day-to-day grind and not wanting to or not having time to work at night on curriculum development was what happened. (Fall 2010)

As we discussed his work, Richard said that he was not sure what teaching social justice through socioscientific issues would look like, but that regardless of his expectations of what he could and could not manage in his quest to act as the activist teacher he wished to be, the ideal was still worth striving for.

CONCLUSIONS

Figure 2 shows a professional belief web specific to Richard.

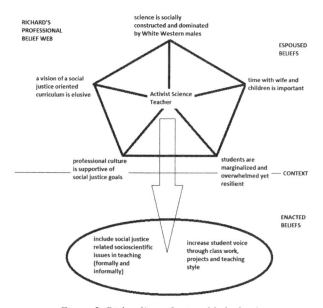

Figure 2. Richard's professional belief web.

It shows the professional identity to which he aspired – activist science teacher – as the centre of his belief web. The other espoused beliefs connected to it did not necessarily always pull equally. At certain points during the four years of the study, different nodes were more important and skewed the web in their direction. For example, during the beginning of his fifth year teaching, as he prepared for interviews for the curriculum leader position, his web skewed toward professional culture. Similarly, once he got the position and felt challenged both professionally by having to deal with new responsibilities and personally in learning that he had a second child on the way, the web was pulled in three directions: PCK, life balance, and professional culture. These pulls in turn affected the trajectory from espoused to enacted beliefs.

Partly because of his participation in this study, Richard was keenly aware of the distance between his espoused and enacted beliefs. This distance framed his experience as he strove toward his goal of becoming an activist science teacher. Richard's experience of becoming an activist science teacher is instructive in that he accomplished a great deal yet still felt frustrated in spite of the moral support he had from his principal and colleagues (both on and off site) and myself. What this research project provided for him was an outside view of his accomplishments as a science teacher. He had included several social justice-related socioscientific issues (and social justice, in general) in his teaching, a fact about which I often had to remind him. The significance of Richard's espoused beliefs compared to his enacted ones is that it is the espoused beliefs which framed his experience of enacting his beliefs.

In moving toward his goal, Richard benefitted from: (1) the support of his colleagues, (2) his sophisticated concept of NOS, and, as time went on, (3) the vision needed to achieve his goals. If we examine Richard's experience, it appears that the type of support that a new teacher with social justice goals needs is not only technical but emotional, too. Beyond the moral support of his colleagues, Richard's experience seems to indicate a need for connection with other new (or experienced) teachers with similar goals (Luft et al., 2007; McCann & Johannessen, 2004).

New science teachers can be supported in various ways: (1) through research that brings together new teachers for periodic discussions, (2) through sustained connections with faculties of education, and (3) in the absence of both of those elements, access to the experiences of other new teachers through case studies such as this one to fulfill their need to know that their experiences are mirrored elsewhere. Above all, it is important for teachers to experience their successes as successes. What may be needed is a personalized, science focused, professional learning program (Kose & Lim, 2011; Luft, Firestone, Wong, Ortega, Adams & Bang, 2011) that responds to the needs and context of particular teachers through self-reflection, peer support and a trusted senior mentor. An individualized professional belief web could be a tool in such a program.

The significance of this case study, then, can be summarized as follows:

- It provides an illustrative example of the experience of being a new teacher struggling to become an activist science teacher.
- It shows the ways in which examining teacher beliefs can be used to analyze that experience.
- It provides an example of how cooperative analysis between a participant and participant-observer can be a form of professional development.

Single case studies of individual teachers can serve as tools for understanding the experiences of others in a detailed and personal way. Richard is one teacher in one context, but through his professional belief web, his case has the potential to inform both future research into and the design of professional development for new science teachers.

REFERENCES

Atwey, B. (2011). Reflections on social justice, race, ethnicity and identity from an ethical perspective. *Cultural Studies of Science Education, 6*, 33-47.

Barrett, S. E., Ford, D., & James, C. (2010). Beyond the lone hero: Providing the necessary supports for new teachers in high needs schools. *Bank Street's Occasional Paper Series, 25*, 67–80.

Barrett, S. E., & Nieswandt, M. (2010). Teaching ethics through socioscientific issues in physics and chemistry: Teacher candidates' beliefs. *Journal of Research in Science Teaching, 47*(4), 380–401.

Barrett, S. E., Solomon, R. P., Singer, J., Portelli, J. P., & Mujawamariya, D. (2009). The hidden curriculum of a teacher induction program. *Canadian Journal of Education, 32*(4), 677–702.

Bianchini, J. A., & Brenner, M. E. (2010). The role of induction in learning to teach toward equity: A study of beginning science and mathematics teachers. *Science Education, 94*(1), 164–195.

Brock, C. H., & Boyd, F. B. (2011). Fostering meaningful middle school literacy earning: Investigating beliefs and practices. *Voices From the Middle, 19*(1), 13–18.

Bryan, L. A. (2003). Nestedness of beliefs: Examining a prospective elementary teacher's belief system about science teaching and learning. *Journal of Research in Science Teaching, 40*, 835–868.

Donnelly, L. A., & Argyle, S. (2011). Teachers' willingness to adopt nature of science activities following a physical science professional development. *Journal of Science Teacher Education, 22*(6), 475–490.

Fletcher, S. S., & Luft, J. A. (2011). Early career secondary science teachers: A longitudinal study of beliefs in relation to field experiences. *Science Education, 95*(6), 1124–1146.

Friedman, A. A., Galligan, H. T., Albano, C. M., & O'Connor, K. (2009). Teacher subcultures of democratic practice amidst the oppression of educational reform. *Journal of Educational Change, 10*, 249–276.

Glesne, C. (2011). *Becoming qualitative researchers: An introduction* (4th ed.). Boston, MA: Pearson/ Allyn & Bacon.

Helms, J. V. (1998). Science and me: Subject matter and identity in secondary school science teachers. *Journal of Research in Science Teaching, 35*(7), 811–834.

Hodson, D. (2003). Time for action: Science education for an alternative future. *International Journal of Science Education, 25*(6), 645–670.

Kane, R., Sandretto, S., & Heath, C. (2002). Telling half the story: A critical review of research on the teaching beliefs and practices of university academics. *Review of Educational Research, 72*(2), 177–228.

Kelly-Jackson, C. P., & Jackson, T. O. (2012). Meeting their fullest potential: The beliefs and teaching of a culturally relevant science teacher. *Creative Education, 2*(4), 408–413.

Kose, B. W., & Lim, E. (2011). Transformative professional learning within schools: Relationship to teachers' beliefs, expertise and teaching. *The Urban Review, 43*(2), 196–216.

Lasky, S. (2005). A sociocultural approach to understanding teacher identity, agency and professional vulnerability in a context of secondary school reform. *Teaching and Teacher Education, 21*(8), 899–916.

Luehmann, A. L. (2007). Identity development as a lens to science teacher preparation. *Science Education, 91*(5), 822–839.

Luft, J. A. (2007). Minding the gap: Needed research on beginning/newly qualified science teachers. [editorial]. *Journal of Research in Science Teaching, 44*(4), 532–537.

Luft, J. A. (2009). Beginning secondary science teachers in different induction programmes: The first year of teaching. *International Journal of Science Education, 31*(17), 2355–2384.

Luft, J. A., Bang, E., & Roehrig, G. H. (2007). Supporting beginning science teachers. *Science Teacher, 74*(5), 24–29.

Luft, J. A., Firestone, J. B., Wong, S. S., Ortega, I., Adams, K., & Bang, E. (2011). Beginning secondary science teacher induction: A two-year mixed methods study. *Journal of Research in Science Teaching, 48*(10), 1199–1224.

Lumpe, A., Czerniak, C., Haney, J., & Beltyukova, S. (2012). Beliefs about teaching science: The relationship between elementary teachers' participation in professional development and student achievement. *International Journal of Science Education, 34*(2), 153–166.

McCann, T. M., & Johannessen, L. R. (2004). Why do new teachers cry? *The Clearing House, 77*(4), 138–145.

Merriam, S. B. (1998). *Qualitative research and case study applications in education.* San Francisco: Jossey-Bass Publishers.

Milner, A. R., Sondergeld, T. A., Demir, A. P., Johnson, C., & Czerniak, C. (2012). Elementary teachers' beliefs about teaching science and classroom practice: An examination of pre/post NCLB testing in science. *Journal of Science Teacher Education, 23*(2), 111–132.

Pinnick, C. L. (2008). Science education for women: Situated cognition, feminist standpoint theory, and the status of women in science. *Science & Education, 17*(10), 1055–1063.

Reeves, J. (2007). Inventing the chartered teacher. *British Journal of Educational Studies, 55*(1), 56–76.

Roth, W.-M. (2010). Activism: A category for theorizing learning. *Canadian Journal of Science, Mathematics and Technology Education, 10*(3), 278–291.

Sachs, J. (2003). *The activist teaching profession.* Buckingham: Open University Press.

Strauss, A., & Corbin, J. (1990). *Basics of qualitative research.* Newbury Park: Sage Publications Inc.

Tesch, R. (1990). *Qualitative research: Analysis types and software tools.* Bristol, PA: Falmer Press.

Theriot, S., & Tice, K. C. (2009). Teachers' knowledge development and change: Untangling beliefs and practices. *Literacy Research and Instruction, 48*(1), 65–75.

Topcu, M. S., Sadler, T. D., & Yilmaz-Tuzun, O. (2010). Preservice science teachers' informal reasoning about socioscientific issues: The influence of issue context. *International Journal of Science Education, 32*(18), 2475–2495.

Wenger, E. (1998). *Communities of practice: Learning, meaning, and identity.* Cambridge: Cambridge University Press.

Yin, R. K. (2009). *Case study research: Design and methods* (4th ed.). Los Angeles: Sage Publications.

Zeidler, d. L., & Sadler, T. D. (2008). Social and ethical issues in science education: A prelude to action. *Science and Education, 17*, 799–803.

AFFILIATION

Sarah Elizabeth Barrett
Faculty of Education
York University

VANASHRI NARGRUND-JOSHI, MEREDITH PARK ROGERS &
HEIDI WIEBKE

EXAMINING SCIENCE TEACHERS' ORIENTATIONS IN AN ERA OF REFORM: THE ROLE OF CONTEXT ON BELIEFS AND PRACTICE

Considering the ever-growing demand to prepare students to be globally competitive in science, it is not surprising that many countries are undergoing policy and curriculum changes regarding science teaching and learning. Because the success of new curricula and standards depends on how teachers carry them out, it is critical to understand their beliefs and practices before forging ahead with reforms (Brickhouse, 1990; Bryan & Abell, 1999; Levitt, 2002; Tobin & McRobbie, 1996). Also, to understand discrepancies between teachers' beliefs and practices, it is important to understand the role social or cultural factors may have on teachers' enactment of their beliefs (Little, 2003).

For this study, we adopted Friedrichsen, van Driel and Abell's (2011) position that teachers hold a set of belief structures, a construct which they refer to as a science teaching orientation. They describe this complex set of beliefs as including, a) what teachers know as the goals and purposes for teaching science, b) how teachers view the practices of science as a discipline, and c) what ideas teachers' hold about science teaching and learning science, which are often shaped by their own experiences as learners of science. Thus when we discuss teachers' views and approaches to teaching science as "an orientation towards science teaching" we are referring to this collective set of beliefs.

This study focuses on India's current experience with curriculum reform. India's reform is to similar efforts in other countries; therefore, a discussion is needed on how other countries are designing, or need to be designing, professional development and teacher education programs to meet science teachers' needs for understanding and implementing their country's science curriculum. A necessary first step in this global discussion is to understand, from teachers' perspectives (Levitt, 2002), their beliefs about what science is, how to teach it, and what learning goals to set for their students. To understand discrepancies between discrepancies between teachers' beliefs about science teaching and learning and their actual practices, science teacher educators, professional developers, and other stakeholders need to understand how these sets of beliefs are developed and how they are enacted in practice. With this information, the various stakeholders can then begin to think of ways to help

R. Evans et al., (Eds.), The Role of Science Teachers' Beliefs in International Classrooms, 165–177.

teachers fulfill the expectations of the curriculum. Accordingly, our study explored the context of India with respect to the first two areas of concern – teachers' beliefs and practices. Our goal therefore, was to identify socio-cultural or contextual factors shaping teachers' belief systems, or their orientations, and the enactment of their orientations. Taking this into consideration, the following research questions were developed to guide our exploration.

1. What orientations do teachers hold?
2. How are teachers' orientations implemented in their practice?
3. What contextual factors appear to shape these orientations and how?

Also because this study was conducted shortly after the introduction of India's new National Curriculum Framework (National Council for Educational Research and Training [NCERT], 2005), it offers insight into how socio-cultural or contextual factors might affect curricular reform at a critical point for intervention.

REVIEW OF LITERATURE

Orientations (or belief sets) are often the best indicators of the how individuals will make decisions throughout their lives (Bandura, 1986) and therefore it is essential to understand teachers' orientations and how these impact their decision-making, including beliefs about a) how students learn, b) how to achieve particular instructional tasks, and c) how to teach particular topics to certain grade levels (Fang, 1996). These belief structures, which result in the formation of their science teaching orientations, are often not explicitly developed until prospective teachers are introduced to learning theory and pedagogy in teacher training programs (Abell, Appleton & Hanuscin, 2010). However, when actually enacting these sets of beliefs into practice, teachers naturally try to merge their newly formed belief systems with their preexisting belief systems, which were formed out of their own learning experiences as students of science. This merging of old and new rarely occurs without difficulties, resulting in what Pajares (1992) described as a "messy construct." Consequently, teachers often end up developing multiple belief systems or orientations towards teaching science (Abell, 2007; Park Rogers, Cross, Gresalfi, Trauth-Nare, & Buck, 2011).

Two themes with respect to factors influencing a teacher's orientation towards teaching science recur in the literature. One refers to external or contextual constraints that teachers believe inhibit them from implementing the reform-minded orientation they know they should have (Beck, Czerniak & Lumpe, 2000; Park Rogers et al., 2011). Contextual constraints include the physical learning environment (e.g., building structure and instructional resources), the human or cultural environment (e.g., student, parent, and administrative expectations), and the political environment (e.g. policies and cultural norms) (Ford, 1992). In addition, teachers face personal constraints such as a lack of understanding of the goals of reform or the nature of science, or limited science experiences as both learners and teachers of science that reflect a reform-based approach (Bryan, 2003; Fetters, Czerniak, Fish & Shawberry, 2002; Lederman, 2007; Levitt, 2002; Putnam & Borko, 2000).

In this chapter, we examine two secondary science teachers' orientations including, how these were reflected in their practices and ways in which their orientations and practices might have been affected by contextual constraints.

METHODS

Context

The site of this study was a private, English-medium secondary school in the state of Maharashtra, India. According to NCERT survey conducted in 2005, student enrollment in Maharashtra at the elementary level (grades 1 to 5) was 6,356,602 but at the secondary level (grades 9 and 10) only 1,315,897. This drastic drop in enrollment could be attributed to fewer secondary schools in rural communities, the traditional societal views that girls do not need to be educated beyond elementary and/or middle grade levels, or to concerns for girls, who might need to travel quite a distance in rural areas to attend secondary schools (Kingdon, 2007). Regardless of this low percentage of children going on to secondary or post-secondary education, there is a heavy emphasis on preparing students at all grade levels to do well so they have the opportunity to attend the most revered universities, those that focus on the mathematics, science, computers, and engineering.

Although the NCF-2005 (NCERT, 2005) document was released nearly a decade ago, it was not until 2012 that textbook revisions were completed. Therefore, when our study took place in 2009, the NCF-2005 was available but the new curricular materials were not. This is a significant issue as most teachers do not read policy documents (or standards) like the NCF-2005 but rely instead on the nationally developed textbooks to tell them what topics to teach and how to teach them.

A major reform of the NCF-2005 is a shift from an objectivist view of teaching and learning science to a constructivist view (Vrasidas, 2000). An objectivist view suggest there is only one truth and way to reach this knowledge; whereas, a constructivist view argues "that the world can never become known in one single way. For constructivists, 'learning is meaning-making'" (Vrasidas, 2000, p. 7). Therefore, the NCF-2005 states that students need to be involved in socially constructing their science knowledge through inquiry-based learning (e.g., questioning, investigation, and scientific argumentation grounded in evidence-based claims).

Participants

Both of the teacher participants for this study, Piya and Shweta (pseudonyms), taught at a private school where all subjects were taught in English, not the native language. In India, schools with no government funding are referred to as private unaided or simply private schools. While facilities are usually better in private schools, teachers usually receive low salaries and are expected to supervise extra-curricular activities for no additional compensation.

At the time of data collection, Piya was in her 13th year as a secondary science teacher. She was responsible for teaching both 9th and 10th grade science with each class averaging 35 students. Piya had completed two Master's degrees, in education and in a physics related science (electronics). In addition to the standard 9th and 10th grade science classes, Piya also taught a specialized environmental science class. The second participant, Shweta, was in her 6th year of teaching and taught 8th and 9th grade science. She also held degrees in education and science, but at the Bachelors level. Each of Shweta's classes averaged 60 students. The majority of both teachers' instruction took place in a lecture-style classroom that included a blackboard and a few reference charts posted on the walls. Once a week, Piya and Shweta also took their students to the school's science laboratory.

Data Sources and Analysis

Data sources for this study included researcher field notes and audio-recordings of five or six classroom lessons for each teacher, two of which included laboratory sessions. Each lecture-style observation lasted approximately 30 minutes and each laboratory observation between one to two hours. In addition to classroom observations, both teachers completed a biographical form prior to participating in an individual interview, which took place after all classroom observations were completed. The interview took approximately 90 minutes to complete and included a modified card-sorting task (see Friedrichsen and Dana, 2003 for a description of this interview technique) to elicit each teacher's set of beliefs regarding a) what they knew about the goals and purposes for teaching science, b) their views about the practices of the discipline of science, and c) their personal beliefs about teaching and learning science. In addition, teacher and student artifacts (e.g., science notebooks and class notes) were collected and reviewed for the purpose of elaborating on the field notes. These notes and artifacts also supported the development of our own[1] context specific scenarios used for the card-sort task in the interview, which we developed following Friedrichsen and Dana's (2003) recommendation that scenarios need to fit specific contexts. Finally, we reviewed the NCF-2005 to determine the goals and purpose for teaching and learning science promoted with the curriculum reform.

The identification of each teacher's science teaching orientation was determined through an analysis involving techniques used in grounded theory (Glaser & Strauss, 1967). These techniques call for the iterative and combined use of interpretative and flexible methods of analysis such as close reading, and open or inductive coding (Benard, 2002; Emerson, Fretz & Shaw, 1995). We implemented this method when analyzing each of our data sources in order to answer our three research questions.

We initiated the first level of coding, or what is referred to as open-coding, with the interview transcripts, and asked questions such as: How are teachers interpreting cards and sorting them? What examples are teachers using while sorting cards?

What goals/purpose are teachers entertaining while sorting cards? What factors do the teachers suggest affect their choices to teach science according to particular scenarios? While reviewing field notes and artifacts, we posed such questions as: How are teachers communicating the science content to students? How are students engaged with the content in class (which could vary for lecture versus lab settings)? What is the teacher emphasizing to the students as to the purpose for learning?

After reaching a point of saturation with open coding for all data sources (i.e., no new codes were emerging regarding the teachers' three sets of beliefs), we then employed an inductive approach to begin grouping related codes for the purpose of developing an axial coding schema. This schema allowed for the formation of categories, which we then used to illustrate the relationship between the teachers' beliefs and practices. Examples of our analytical categories included: emphasis on use of a textbook when teaching, teacher relaying or dictating information to the students, teacher perceived constraints on her teaching, and expectations for covering the syllabus. These and other categories are discussed in our findings, which are framed according to the three research questions.

FINDINGS

Identifying Teachers' Orientations towards Teaching Science

Piya She viewed the main purpose for her teaching and her students' learning of science as learning a correct scientific understanding of the concepts. For example, she explained, "Students should develop a scientific attitude and learn to think logically. By developing these qualities they can stay away from superstitions" (Piya, Interview). In addition, she placed a strong emphasis on learning the science content included in the 10th grade board exam, which is cumulative from 8th to 10th grade, so teaching for test preparation is viewed as critical during these grade levels. The following example of this "test prep teaching" was observed during a lesson on classification.

> Piya asked students to open the textbook and explain different ways in which plants and animals utilize nutrients. During this discussion she asked students to refer to their textbook frequently. After the discussion she asked the students, "so what criteria have we discussed so far?" Students replied, "nutrition", "biomolecules", and "nitrogen fixation." She told students to underline the important words production and storage of specific type of chemicals in the textbook because they'll need this for the exam. (Field Notes, 07/08/2009)

Piya explained her reasons for teaching students in this way:

> Underlining is important because during board exam it is very important to write answers in a point wise format . . . if I do not teach them this way then they will write lengthy answers which is not acceptable. (Interview)

Although teaching for this purpose was the focus of much of Piya's instruction, she did occasionally try to connect the content she was teaching to some aspect of the students' everyday lives. She explained that she felt these sorts of connections were important to make because "otherwise they think whatever is given in the textbook is different than what they experience in their lives and they start losing interest. Once we make these connections, students start thinking and understand that we are learning something that is connected to our lives" (Piya, Interview).

Piya demonstrated a mixed set of beliefs about the goals and purposes for teaching science. On one hand she felt her primary goal was to teach them the content to be successful on the test, but on the other hand she believed it was important to make connections to the students' everyday lives so they would develop an appreciation for science. Regardless of her beliefs about either goal, she believed that science learning meant developing an understanding of one standard explanation of a science phenomenon.

Another set of beliefs shaping Piya's orientation towards teaching science included her views of science practices. Piya wanted her students to understand how scientific knowledge develops because of scientists' persistent efforts, which she expressed when sharing her frustration about the removal of Darwin's contribution from the science textbook. She said,

> Students should understand who has made original discoveries and they should also understand how scientists have worked constantly over years before coming up their discoveries. [They need to know] it did not happen in one night. (Interview)

At the same time, she believes that school science needs to be learned through an objective rather than subjective lens, stating, "I prefer asking questions that have fixed outcome because in science there is a correct answer and they can give examples of their choice while answering but their answer should be based on textbook content" (Interview). Conversely, she agreed that students need to develop the process skills of inquiry science (e.g., making observations and inferences, addressing questions empirically, etc.), suggesting the belief that science learning should involve investigation. This contradiction in her beliefs about the practice of science and how students should learn the content in school science suggests Piya shifted her orientations between objectivist and constructivist perspectives depending on what set of beliefs within the structure of her orientation she was discussing.

Lastly, Piya's beliefs about how she should teach and students should learn science remained fairly consistent throughout our discussions with her. We also observed that she relied heavily on the textbook as the primary source for content knowledge, which she confirmed with the following comment.

> I am more textbook orientated because the tests are more than 10% objective types of questions (e.g., fill in the blanks). So having students read the textbook several times helps them to start to recognize the words so they'll recognize them and remember what word goes in the right spot. (Interview)

Shweta Contrary to Piya, Shweta viewed the purpose for teaching science as only preparing her students for the 10th grade board exam. In India, a career in the sciences is most highly regarded, and the 10th grade board exam determines what academic path students take for the remaining two years of secondary schooling in preparation for university studies. Therefore, to have a chance for a science related career, students need to do well on the science portion of this exam. Throughout the interview Shweta frequently used expressions such "feeding data to kids," "sticking to the textbook," or "sticking to the board exam question paper pattern." When asked about using science magazines as a resource, she replied,

> because of time limitations we cannot go beyond what the textbook says. If students know some information about the topic we are studying and want to share, they can post it to the board outside of the classroom. Then whoever is interested in reading can read it on their time (Interview).

Not surprisingly, we often observed Shweta dictating notes to her students or having them copy from the book.

Regarding the practices of science, Shweta believed the discipline of science is objective, which translates to focusing on teaching students that there is one right answer they need to learn. As she stated, "we just read [about the concept] in the textbook chapter and move on. It is not really important from the point of view of preparing for the exam that the students know how ideas came to be known" (Shweta, Interview). This belief was also evident in her description of how she planned her instruction. She explained that she had a specific formula for students to follow when answering questions from the textbook and writing lab reports, which was the only structure for providing answers that would be accepted on the test.

Lastly, we observed that Shweta's teaching involved lecturing to students from the textbook and defining terms according to the textbook. When asked how she prepared for her lessons, she stated,

> I define points. I go through the textbook points to be sure I can explain them as per the textbook. That means I don't go into detail if it is not mentioned in the textbook. This process also helps me with framing questions properly. (Interview).

An example of her focus on the text was our observation of Shweta telling students which pages and line numbers to highlight in their textbook so they would have the definitions to practice.

Our analysis of the teachers' three integrated sets of beliefs identified Shweta as having an objectivist orientation and Piya as shifting between objectivism and constructivism on the continuum (Vrasidas, 2000), depending on which set of beliefs she was discussing. For example, she believed students need to understand specific conceptions of science, an objectivist stance, but to make sense of concepts they also needed to connect them to their everyday lives, evidence of a more constructivist orientation. However, she also felt compelled to ensure her students were learning to answer questions in ways specifically expected for the 10th grade board exam, again an objectivist view. This shifting between orientations is perhaps what Pajares (1992) was referring to when he described educational beliefs as being a "messy construct."

Translation of Teachers' Orientations into Practice

Piya   In addition to her lecture time, each week Piya was assigned 60 minutes of laboratory time with her students. In our observations it was evident that Piya's shifting of orientation also influenced her approach to laboratory experiences. For example, her objectivist orientation was evoked by the need for students to complete the labs in a manner that confirmed what they had previously learned in the classroom. However, her constructivist orientation would sometimes emerge, as when she asked students to consider why the results of their investigations were different from what they expected. We observed an example of this when Piya's students were preparing slides of a fungus:

> The students were maintaining two science notebooks – one where they wrote their observations of the slide under microscope, and a second where they copied the observations and figures given at the end of the textbook journal. (Field Notes, 07/13/2009)

For Piya, the laboratory offered students the opportunity to construct meaning of concepts through their own observations, but she also wanted to be sure they recorded the correct information for the test, so students kept track of both.

When given scenarios that suggested open inquiry experiences for students, she rejected them, saying, "some bright students might be able to perform activities without anyone's help but there are other students who might not be able handle things responsibly" (Interview). She further explained that for the "average student, demonstrations and clear instructions work best and for this to occur the class needs to well-managed" (Interview). When probed about her notion of "well-managed," Piya explained, "I tell them not to ask me questions when I am teaching because then I forget what I have already taught. I really prefer that they wait to ask questions at the end of the lesson" (Interview). This preference for a classroom structure in which students listened to her lecture and took notes as she directed, with only a few minutes at the end of the class to ask clarifying questions revealed an objectivist orientation. When asked how she learned to teach in this manner, Piya explained that it was how she had learned science.

Shweta   During the card-sort interview Shweta was reluctant to choose cards that suggested open-inquiry types of investigations for students although unlike Piya she did not reject them outright. When asked about having students building models to explain their thinking, she said, "I would do this activity as a review for a test if I get time," but when probed further she explained she would expect her students' explanations of the model to mirror the explanation in the textbook. Comments such as this demonstrate Shweta's objectivist science teaching orientation. She believed only the textbook provided the information students needed to learn.

Regarding the writing of laboratory reports, Shweta said, "whatever [the students] write in their journal should be copied using the outline in the back of the journal" (Interview). After asking Shweta to elaborate, we learned that the national textbook series, which most schools in India use, comes with a student notebook that provides the structure for a laboratory report. Although Shweta said it was important that students understand the process of writing laboratory reports, she only required them to copy the report from the student notebook into their class notebook. For her this was also another way to teach students to use the formula for answering questions expected on the board exam.

One difference we observed in Shweta's teaching compared to Piya's was her use of textbook provided charts and diagrams during her lectures, which she said helped explain concepts in the textbook. She believed that if the students saw a pictorial representation of a concept as well as read about it, they would understand it better. However, unlike Piya, Shweta asked fewer questions of her students in class. She believed it was her job to relay information and not pose questions to students. Instead, students responded to questions from the textbook as practice for board exams. From these findings, it was evident Shweta's objectivist orientation translated directly into her practice, and the need to prepare students for success on the exam validated her teaching in this way.

Contextual Factors Shaping Teachers' Orientations

Our final research question focused on identifying contextual factors that Piya and Shweta identified, or we observed, as influencing their orientation towards teaching and its translation into practice. We identified two factors, one common to both, and one specific to Piya.

The shared factor was the need to prepare students to learn the accepted method for answering questions on the board exam. As Shweta said, "they are the most important. Students' careers depend on them, so we have to prepare them and to do so means preparing them to answer questions according to the board exam pattern" (Interview). Although perhaps not as deliberate as Shweta, Piya also sometimes gave her students specific directions on how to frame answers to certain questions.

The factor that only Piya discussed was how her own experience with learning science influenced her belief about how to structure her teaching. As previously stated, Piya followed a format of presenting information while students listened, took notes, and asked clarifying questions near the end of class. Having learned science in this way herself, Piya had no other model of instruction, so she simply repeated her own learning experience.

CONCLUSIONS

We identified Piya as having a shifting orientation, with some of her beliefs aligning with an objectivist view towards the teaching of science and others aligning with a constructivist view towards the practices of science. These contrasting sets of beliefs resulted in her having different goals for how she wished her students could learn science and the reality of how she taught them. This phenomenon of holding multiple orientations, or shifting between orientations, which has been found to be common among teachers, adds confusion to understanding the complexity of the relationship between beliefs and practice (Friedrichsen et al., 2011). For the most part, Piya's shifting orientation was the result of tension among what she believed her students should learn about science to understand their world, her own experiences with learning science, and the pressure to prepare students for the 10th grade board exam.

Shweta on the other hand was identified as consistently holding an objectivist orientation, as each set of her beliefs endorsed a single truth to be learned and a single method for learning that included practicing the format required for the board exam. In Shweta's case, only the contextual factor of test preparation shaped her orientation.

The discrepancy between Piya and Shweta offers some insight into the debate in the literature on the whether beliefs influence practice (Mansour, 2009). Our study lends weight to both sides of the debate. For Piya, only certain aspects of her mixed orientation aligned with what we witnessed in her teaching; whereas Shweta's teaching supported the notion that practice can directly reflect beliefs (Fang, 1996; Park Rogers et al., 2011). For both teachers, the contextual factor of the 10th grade board exam, which resulted in teaching to the test and the textbook being seen as the authority (Anderson, 2002) for what the students needed to know and how they needed to learn it. These findings support the idea that until the pressure of exam preparation are removed, such dependency on textbooks and didactic instruction will remain (Abrams, Pedulla & Madaus, 2003).

Based on this research, we recommend that high priority be placed on the following issues: a) systemic change regarding the purpose of school science, b) assessment policy and design, and c) professional development programs. Discussion regarding systemic changes in what is the purpose of school science must include all stakeholders, including government curriculum agencies, school administrators, teachers and teacher educators, parents, students, and university officials. A vital component of this conversation must be consideration of the central role high-stakes tests play in determining students' post-secondary choices. Stakeholders need to come to an agreement concerning the goal for learning science in the early years; is it to track students into specific careers or to develop their scientific literacy and critical thinking skills so they can develop an understanding of the natural world? Right now, the former seems to be the culturally accepted norm in India, but the latter is the stated goal of the NCF-2005.

Along with reaching consensus on the primary goal for learning science, assessments also need to be developed to reflect this purpose. According to the NCF-2005, emphasis should be on understanding the practices of science as much as learning the core disciplinary ideas. While Piya agreed with this goal personally, the pressure of high-stakes assessments directed her to teach in a way that contradicted the constructivist aspect of her orientation, such as when we observed that she had her students keep two lab notebooks – one based on their own observations and one that mirrored the textbook. Creating new assessments can be a long and arduous process, but matching assessment to reform goals is a necessary next step to accurately measure their impact on student learning (National Research Council, 2001).

Lastly, as our study shows, teachers' orientations are indeed complex, and many factors can influence their development and relationship to teachers' practices (Mansour, 2009). Therefore, the final implication from this study is the importance of effectively delivering the message of the reform to all stakeholders. In India, the government developed textbook series used in most schools is logically the first place to begin communicating the message of the NCF-2005. However, our study occurred four years after the NCF-2005 was initially introduced, yet new textbooks had still not been developed. This situation needs to be rectified as soon as possible because, as we learned from Piya and Shweta, teachers see the textbook as the source for learning about science in school. Until this mindset changes, the textbook will continue to be teachers' primary resource for deciding what to teach and how to teach it.

Consideration also needs to be given to professional development that targets teachers' orientations toward teaching science. At the time of this study, it was apparent teachers might have received a copy of the NCF-2005, but limited professional development was provided to help them understand it. Perhaps representative of many teachers with confirmed objectivist orientations, who believe they are fulfilling the curriculum framework goals, Shweta said, "I remember reading [the NCF-2005] when I first got it but now I don't really remember what is in it. However, I am sure I am following all the teaching instructions mentioned in that document" (Interview). Clearly, teachers need the opportunity to explicitly identify their beliefs about and orientations for teaching science and assess how well they align with the goals of the curriculum reform, which should be the first mission of professional development. Following this explicit identification and alignment process, professional developers can then deliver a consistent message that will support teachers in developing and sustaining orientations compatible with reform goals as well as offer strategies for enacting their reform-minded orientations into practice (Capps, Crawford and Constas, 2012; Park Rogers et al., 2007).

NOTE

[1] See Nargund-Joshi and Rogers (in review) for a description of our context-specific card-sort interview.

REFERENCES

Abell, S. K., Appleton, K., & Hanuscin, D. L. (2010). *Designing and teaching the elementary science methods course*. New York, NY: Routledge.

Abell, S. K. (2007). Research on science teacher knowledge. In S. K. Abell & N. G. Lederman (Eds.), *Handbook of research on science education* (pp. 1105–1150). Mahwah, NJ: Lawrence Erlbaum Associates, Inc., Publishers.

Abrams, L. M., Pedulla, J. J., & Madaus, G. F. (2003). Views from the classroom: Teachers' opinions of statewide testing programs. *Theory into Practice, 42*(1), 18–29.

Anderson, R. D. (2002). Reforming science teaching: What research says about inquiry. *Journal of Science Teacher Education, 13*(1), 1–12.

Bandura, A. (1986). *Social foundations of thought and action: A social cognitive theory*. Englewood Cliffs, NJ: Prentice Hall.

Beck, J., Czerniak, C. M., & Lumpe, A. T. (2000). An exploratory study of teachers' beliefs regarding the implementation of constructivism in their classrooms. *Journal of Science Teacher Education, 11*(4), 323–343.

Bernard, H. R. (2002). *Research methods in anthropology: Qualitative and quantitative approaches* (5th ed.). Walnut Creek, CA: AltaMira Press.

Brickhouse, N. W. (1990). Teachers' beliefs about the nature of science and their relationships to classroom practice. *Journal of Teacher Education, 41*(3), 53–62.

Bryan, L. A. (2003). Nestedness of beliefs: Examining a prospective elementary teacher's belief system about science teaching and learning. *Journal of Research in Science Teaching, 40*(9), 835–868.

Bryan, L., & Abell, S. (1999). The development of professional knowledge in learning to teach elementary science. *Journal of Research in Science Teaching, 36*(2), 121–140.

Capps, D. K., Crawford, B. A., & Constas, M. A. (2012). A review of empirical literature on inquiry professional development: Alignment with best practices and a critique of the findings. *Journal of Science Teacher Education, 23*(3), 291–318.

Emerson, R. M., Fretz, R. I., & Shaw, L. L. (1995). *Writing ethnographic fieldnotes*. Chicago: University of Chicago Press.

Fang, Z. (1996). A review of research on teacher beliefs and practices. *Educational Research, 38*(1), 47–65.

Fetters, M. K., Czerniak, C. M., Fish, L., & Shawberry, J. (2002). Confronting, challenging, and changing teachers' beliefs: Implications from a local systemic change professional development program. *Journal of Science Teacher Education, 13*(2), 101–130.

Ford, M. E. (1992). *Motivating humans: Goals, emotions, and personal agency beliefs*. Newbury Park, CA: Sage Publications.

Friedrichsen, P. M., & Dana, T. M. (2003). Using a card-sorting task to elicit and clarify science-teaching orientations. *Journal of Science Teacher Education, 14*(4), 291–309.

Friedrichsen, P., van Driel, J. H., & Abell, S. K. (2011). Taking a closer look at science teaching orientations. *Science Education, 95*(2), 358–376.

Glaser, B. G., & Strauss, A. L. (1967). *The discovery of grounded theory: Strategies for qualitative researcher*. Piscataway, NJ: AdlineTransaction.

Kingdon, G. G. (2007). The progress of school education in India. *Oxford Review of Economic Policy, 23*(2), 168–195.

Lederman, N. G. (2007). Nature of science: Past, present, and future. In S. K. Abell & N. G. Lederman (Eds.), *Handbook of research in science education* (pp. 831–879). Mahwah, NJ: Lawrence Erlbaum Associates, Inc., Publishers.

Levitt, K. E. (2002). An analysis of elementary teachers' beliefs regarding the teaching and learning of science. *Science Education, 86*(1), 1–22.

Little, J. (2003). Inside teacher community: Representations of classroom practice. *The Teachers College Record, 105*(6), 913–945.

Lumpe, A. T., Haney, J. J., & Czerniak, C. M. (2000). Assessing teachers' beliefs about their science teaching context. *Journal of Research in Science Teaching, 37*(3), 275–292.

Mansour, N. (2009). Science teachers' beliefs and practices: Issues, implications and research agenda. *International Journal of Environmental and Science Education, 4*(1), 25–48.

Nargund-Joshi, V., & Rogers, M. A. P. (In Review). Modifying a card sort activity for the purpose of understanding Indian teachers' orientations for teaching science. *Journal of Science Teacher Education.*

National Research Council. (2001). *Knowing what students know: The science and design of educational assessment.* Washington, D.C.: National Academy Press.

National Council of Educational Research and Teaching. (2005). *National curriculum framework.* Retrieved October 1, 2013, from http://www.ncert.nic.in/html/pdf/schoolcurriculum/framework05/prelims.pdf

Pajares, M. F. (1992). Teachers' beliefs and educational research: Cleaning up a messy construct. *Review of Educational Research, 62*(3), 307–332.

Park Rogers, M. Abell, S., Lannin, J., Wang, C., Musikul, K., Barker, D., & Dingman, S. (2007). Effective professional development in science and mathematics education: Teachers' and facilitators' views. *International Journal of Science and Mathematics Education, 5,* 507–532.

Park Rogers, M. A. Cross, D. I., Gresalfi, M. S., Trauth-Nare, A. E., & Buck, G. A. (2011). First year implementation of a project-based learning approach: Addressing teachers' orientations and professional experience. *International Journal of Science and Mathematics Education, 9*(4), 893–917.

Putnam, R. & Borko, H. (2000). What do new views of knowledge and thinking have to say about research on teacher learning? *Educational Researcher, 29*(1), 4–15.

Tobin, K., & McRobbie, C. J. (1996). Cultural myths as constraints to the enacted science curriculum. *Science Education, 80*(2), 223–241.

Vrasidas, C. (2000). Constructivism versus objectivism: Implications for interaction, course design, and evaluation in distance education. *International Journal of Educational Telecommunications, 6*(4), 339–362.

AFFILIATION

Vanashri Nargrund-Joshi
Department of Elementary and Secondary Education - Science Education
School of Education
New Jersey City University - Jersey City

Meredith Park Rogers
Department of Curriculum and Instruction - Science Education
School of Education
Indiana University - Bloomington

Heidi Wiebke
Department of Curriculum and Instruction - Science Education
School of Education
Indiana University - Bloomington

KADIR DEMIR & CHAD D. ELLETT

SCIENCE TEACHER SELF-EFFICACY BELIEFS, CHANGE PROCESSES, AND PROFESSIONAL DEVELOPMENT

INTRODUCTION

This chapter conceptually integrates three frameworks: teacher self-efficacy beliefs, teacher change processes, and teacher professional development as a means of improving teaching and learning in science. The self-efficacy discussion is grounded in the theoretical conceptions of Bandura's (1997) description of the sources of self-efficacy beliefs (enactive mastery, vicarious experience, verbal persuasion, and physiological and affective states) as described more fully in Chapters 1, 2, 7, 12, and 13. As detailed in these chapters, strengthening self-efficacy beliefs is important to teacher change *processes*, which in turn, provide a basis for understanding and explaining successful and unsuccessful efforts in teaching science

We discuss teacher change *processes* using the Concerns-Based Adoption Model (CBAM) based on the work of Hall and Hord (1987, 2001). Subsequently, the conceptual and empirical linkage between strengthening teacher self-efficacy beliefs and CBAM change *processes* is described. Finally, we describe and link professional development to self-efficacy theory and CBAM change processes. Relevant empirical studies document the viability of linking these three frameworks as a means of improving teaching and learning in science. Hence, explicating the linkages between self-efficacy beliefs, the change process, and professional development, in view of student learning and achievement outcomes, is the goal of this chapter.

EXPLICATING LINKAGES BETWEEN TEACHER SELF-EFFICACY, CHANGE PROCESSES, AND TEACHER PROFESSIONAL DEVELOPMENT

Teacher Self-efficacy

There is an historical and rather rich theory and empirical base to support the importance of self-efficacy beliefs in human agency. This section includes a brief description of core concepts comprising self-efficacy theory. Hence, based on Bandura's (1997), account of self-efficacy beliefs and how they influence people's action through what they choose to pursue, it is important to gain a better understanding of how people persevere when they face challenges and adversities. Thus, people with strong self-efficacy beliefs about their capabilities to accomplish

R. Evans et al., (Eds.), The Role of Science Teachers' Beliefs in International Classrooms, 179–190.
© *2014 Sense Publishers. All rights reserved.*

challenges associated with new innovations and changes are more likely to accept and engage and persist in these challenges than those with weak self-efficacy beliefs. As well, successfully adapting to elements of new innovations serves to strengthen an individual's self-efficacy beliefs. Thus, teachers with strong self-efficacy beliefs about their capabilities to design and implement inquiry-based teaching and learning could, at the same time, have weak beliefs about student achievement outcomes.

According to Morris (2004), numerous studies have shown that human behavior can often be better predicted by individuals' beliefs in their capabilities than by their actual capabilities. Educational studies have examined self-efficacy in view of college and career choices, teaching and student outcomes, the efficacy beliefs of teachers, academic performance and achievement (Pajares, 1997). Self efficacy has also been studied from many other perspectives in education such as: beginning science teachers and efficacy beliefs (Chester & Beaudin, 1996; Smolleck & Mongan, 2011; Woolfolk Hoy, & Burke-Spero, 2005); preservice education of teachers (Fitchett, Startker, & Salyers, 2012); beliefs about science teachers' teaching contexts (Lumpe, Haney, & Czerniak, 2000); teacher efficacy and perceived success and mainstreaming students (Brownell & Pajares, 1999); inquiry-based science teacher education and teacher self-efficacy (Evans, 2011; Gado, Ferguson, & van't Hooft, 2006); changes in science teachers' self-efficacy beliefs and science teaching and learning environments (Andersen, Dragsted, Evans, & Sorensen, 2004; Roberts, Henson, Tharp, & Moreno, 2001); professional development and teacher efficacy (Ross & Bruce, 2007); self-efficacy and personal science teaching (Khourey-Bowers & Simonis, 2004); and science content and its effect on science teaching efficacy (Lakshmanan, Heath, Perlmuter, & Elder, 2011).

In a perusal of the literature on self-efficacy beliefs there has been a clear interest in self-efficacy and teachers, teacher education, and teaching, including the teaching of science. To understand the self-efficacy beliefs of science teachers (or others) it seems necessary to understand variables that strength or weaken self-efficacy beliefs. According to Bandura (1997), there are four primary sources that serve to develop and strengthen self-efficacy beliefs: 1) enactive mastery; 2) vicarious experience; 3) verbal persuasion; and 4) physiological and affective states (these are detailed in Chapters 1, 2, 7, 12, and 13).

These four sources of self-efficacy beliefs represent a highly interrelated system that reflects a rich history of various learning theories in psychology. These sources vary in strength and influence depending upon existing capabilities, social interactions with others (including modeling), the difficulty of tasks, self-reflection, and host of other variables. The four sources of self-efficacy are not independent entities and they may be evident in different combinations depending on the task in hand. For example, a student's enactive mastery of a problem in science might easily be accompanied by a teacher's verbal persuasion.

As conceptualized in this chapter, the sources of self-efficacy beliefs are key components for designing and implementing effective professional development programs for teachers (see Chapter 12). Bandura (1997) conceptualizes self-efficacy beliefs as a key element of human learning and performance across any performance domain (e.g., cooking, driving, playing golf, science teaching). Regardless of the task performed, differences in performance are primarily due to variation in the strength of self-efficacy beliefs. These beliefs also explain variation in performance and effectiveness among individuals within a performance domain. Thus, differences among science teachers in the quality of teaching can be explained within self-efficacy theory by different strengths of these beliefs. From the self-efficacy perspective, strengthening teacher self-efficacy beliefs is a primary goal of professional development. As well, strengthening self-efficacy beliefs of teachers enhances changes in practice.

Teacher Change and Teacher Change Processes

Teacher change. Change is both a noun and a verb. Thus, and for example, a difference in how teachers vary their practices as a result of professional development can be considered a teacher change. On the other hand, teacher change can be understood as a process that teachers experience in response to new ideas and practices. Teacher change is usually associated with some innovation frequently linked to professional development aimed at bringing about modifications in teachers' teaching and learning practices.

Teacher change has been largely discussed and researched as a change event or response to an innovation rather than a process. In this line of inquiry, of interest are teaching practices, teacher attitudes and beliefs, and, ultimately, learning outcomes of students. Yet, what constitutes teacher change is broad and fluctuates depending on the perspective one has embraced. For example, some scholars look to both the organization and the teacher to explicate the dynamics that affect the implementation of change (e.g., Fullan, 2007).

Loucks-Horsley, Love, Stiles, Mundry, and Hewson (2003) defined teacher change from both an individual and an organizational perspective, Richardson (1990), on the other hand, defined teacher change, "as teachers doing something that others are suggesting they do. Thus, the change is deemed as good or appropriate, and resistance is viewed as bad or inappropriate" (p. 11). From a broader perspective, Smith, Hofer, Gillespie, Solomon, and Rowe (2003) delineated teacher change as differences in thinking and acting on and off the topic. According to Smith et al. change *on the topic* (e.g., content focus and new teaching practices) comprises increased knowledge about the topic and reported action taken to address learner persistence in the classroom, in the program, or in the field. Change *off the topic* involves increased awareness of the field of adult education, increased confidence in teaching, decreased feelings of isolation, or increased use of a new teaching technique. In their definition of change, Clarke and Hollingsworth (2002) emphasized teachers' *enactment* –the translation of the teacher's knowledge or beliefs into action— and reflection in the professional context.

In addition to the views and perspectives described above, a variety of models have been developed to help improve the understanding of change in individuals and organizations. Each model has steps or characteristics that explicate the conceptual complexity of change and how elements of change are interconnected. The context in which these models play out also varies from one model to the next. For example, from a teacher professional perspective and a rather linear approach, Guskey's (2002) change model emphasizes that change in practice leads to change in students that then leads to change in beliefs. From a cyclical conception, Huberman (1983, 1995) argues that changes in teacher beliefs lead to changes in teaching practices that bring about changes in student learning that bring about further changes in teacher practices that result in additional changes in beliefs and so on. The consensus is that although the literature provides guidance, there is no set recipe for change. Fullan (1992) depicted teacher change as a, "technically simple and socially complex" (p. 109) phenomena. Yet, developing a definitive theory of change is, by definition, not possible. "It is a theoretical and empirical impossibility to generate a theory [of change] that applies to all situations" (Fullan, 1999, p. 21).

Teacher change processes. Our working definition of *teacher change processes* refers to internal psychological events that teachers experience in response to an innovation. Innovations include new practices, policies, knowledge, and/or activities comprising new learning and altered professional perspectives and dispositions. Thus, a *change event* (e.g., new curriculum requirement such as Next Generation Science Standards) is not the same as a *change process*. Our working definition is consistent with elements of change reflected in the Stages of Concern (SoC) and Levels of Use (LoU) comprising the Concerns-Based Adaption Model [CBAM (Hall & Hord, 1987). In agreement with Loucks-Horsley et al. (2003), we view the change processes as progressive, occurring through "active engagement with new ideas, understandings, and real-life experiences" (p. 39).

CBAM, then, provides a "theoretical sequence of stages...which indicate that teachers [or other individuals] pass through a series of stages of concern when they are attempting or expected to use new innovative materials" (Malone, 1984, p. 756). Consequently, CBAM assumes that in any setting employing an innovation, as the change process evolves, individuals' feelings about this innovation will change. While originally designed as a model for understanding change in professional development programs for teachers, the CBAM has been more widely used as a model for understanding change processes in organizations, efforts to evaluate systemic change (Driksen & Tharp, 1997), and teacher concerns regarding implementation of a new science and mathematics curricula (Ellett, Demir, Martin-Hansen, Awong-Taylor, & Vandergrift, 2012; James & Hall, 1981; Christou, Eliophotou-Menon, & Philippou, 2004).

The CBAM includes several perspectives for observing the change process (Hall & Hord, 1987, 2001). These include, but are not limited to: (a) change is a process, not an event or a product; (b) change is not easy and takes time to institute; (c) individuals are the essence of the change process; (d) institutions will not change unless their members change; (e) the change process is a personal experience; (f) individuals' perceptions of change strongly influence the outcome; and (g) individuals' progress through predictable stages regarding their emotions and capabilities related to reforms.

The CBAM includes three key tools used to collect relevant data: SoC, LoU, and IC. The most important tool in the model is the SoC questionnaire, which is used to measure teachers' concerns about an innovation they are expected to implement (Hall & Hord, 2001). The SoC dimension of the CBAM provides the means for assessing the concerns of individuals involved in change. The seven stages of concern are: Awareness (two levels), Informational, Personal Management, Consequences, Collaboration, and Refocusing. These seven stages of concern can further be grouped into three dimensions according to developmental nature of concerns— self, task, and impact. A brief description of each stage is given as follows.

Self Dimension: Early stage of change

Awareness (stage 0):	Teachers are less informed about the innovation and are not attentive towards taking any action.
Example:	I am not concerned about it (innovation).
Informational (stage 1):	Teachers indicate an interest in the innovation and the prerequisites for its implementation. Teachers are inclined to learn more about the particular innovation.
Example:	I would like to know more about it.
Personal (stage 2):	Teachers sense uncertainty about their role in the process of acceptance of the innovation. They may exhibit concerns about the hypothetical benefits of implementation and the possible personal and current program conflicts that might occur as a result of adoption, concerned about their own time limitations and the changes they will be expected to make.
Example:	How will using it affect me?
Task Dimension:	Beginning of use of a reform and early period of use
Management (stage 3):	Teachers are focused on the processes and task of using the innovation, e.g., time, management, efficiency, and constraints.
Example:	I seem to be spending all my time getting material ready.
	Impact Dimension
C *onsequences (stage 4)*:	Teachers concentrate on the outcome of the innovation /change in practice to see if it had any effects on student learning. Positive outcomes are most likely to reinforce teachers' use of innovation.
Example:	How is my use affecting kids?

Collaboration (stage 5):	Teachers are enticed to communicate with their colleagues about relating what they are doing to what their colleagues are doing.
Example:	I am concerned about relating what I am doing with what other teachers are doing.
Refocusing (stage 6):	Teachers carry out an evaluation of the innovation and construct recommendations for sustained practices or deliberate alternative ideas.
Example:	I have some ideas about something that would work even better.

Personal concerns make use of a powerful influence on the implementation of reforms and determine the type of assistance that teachers may need in the adoption process. The results of previous studies show that the perceptions of those involved in reforms are of major importance for the success of the reform process (Fullan, 2007; Senger, 1999). The meaning that teachers attach to an innovation affects their reactions to the innovation and the possible problems associated with these reactions. Thus, it is useful for administrators and educators to have a picture of teachers' concerns, both before and during the implementation phase of a reform (Fullan, 1999, 2007).

Teacher Professional Development.

Teaching is a dynamic pursuit that requires proficiency from teachers as new knowledge about science teaching and learning comes to light. Science teachers need to remain aligned with this knowledge base and use it to constantly process their conceptual and pedagogical abilities and understandings. Professional development activities are believed to enhance the quality of teaching and learning by helping teachers augment and advance their subject matter knowledge, develop new teaching and learning practices, sharpen their existing skills, and engage them in professional growth as teachers (Borko, 2004; Davis, Petish, & Smithey, 2006; Richardson & Placier, 2001). In this chapter we describe how professional development is linked to change processes and teacher self-efficacy beliefs.

The section that follows presents our conceptual framework for thinking about the integration and influences of teacher self-efficacy beliefs, change processes, and professional development programs. Included in this section are examples of the kinds of research studies that might be used to test basic assumptions of the conceptual framework.

A CONCEPTUAL FRAMEWORK LINKING TEACHER SELF-EFFICACY BELIEFS, CHANGE PROCESSES, AND PROFESSIONAL DEVELOPMENT

An emerging body of research documents the relationships between teacher self-efficacy and teacher professional development (e.g., Khourey-Bowers & Simonis,

2004; Lakshmanan et al., 2011; Martin, McCaughtry, McCaughtry, & Cothran, 2008; Posnanski, 2002; Powell-Moman, & Brown-Schild, 2011; Roberts et al., 2001; Ross & Bruce, 2007; Sinclair, Naizer, & Ledbetter, 2011). In addition, there have been calls for including a stronger focus on teacher self-efficacy in research on teacher professional development programs (Ballone-Hartzell & Czerniak, 2001; Bray-Clark & Bates, 2003). Some studies have been completed to examine the relationship between teacher change and teacher professional development (e.g., Boyle, Lamprianou, & Boyle, 2005; Guskey, 2002; Smith & Gillespie, 2007). Fewer studies of teacher change *processes* and teacher professional development have been completed to date (e.g., Greensfeld & Elkad-Lehman, 2007). Our review of the literature examining the role of teachers' self-efficacy beliefs and teacher change *processes* appears to be limited to a single study by McKinney, Sexton, and Meyerson (1999).

To help advance the field in areas where limited research has occurred, we developed a comprehensive conceptual framework that depicts interrelationships among teacher self-efficacy beliefs, teacher change *processes,* and professional development. In theory, we believe that the conceptual framework links these constructs to the enhancement of teaching and learning practices and, ultimately, student achievement outcomes. The conceptual framework is shown in Figure 1.

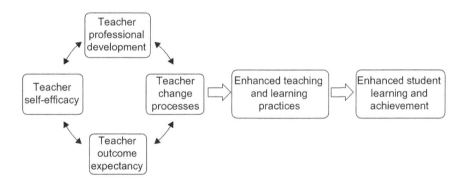

We developed this framework to highlight the importance of linking and integrating teachers' personal beliefs and change *processes* with professional development. There are several features and assumptions about the constructs included in the conceptual framework. For example, the left side of the framework shows reciprocal interactions among teacher professional development, teacher self-efficacy beliefs (including teacher outcome expectancy), and teacher change *processes*. The bi-directional arrows linking these variables are consistent with the assumptions of models of reciprocal determinism such as Bandura's (1997) model of triadic reciprocal causation. In our framework, teacher professional development represents the external learning environment; self-efficacy beliefs, outcome expectancies, and change processes are teacher personal factors (cognitive, affective, and psychological events); and enhanced teaching and learning practices represent behavior.

The framework assumption of reciprocal causation suggests, for example, that changes in teachers' self-efficacy beliefs are causally linked to professional development activities. In turn, these changes brought on by teacher professional development serve to strengthen (or weaken) teachers' self-efficacy beliefs. In our model, teacher change *processes* are influenced by teacher professional development activities and teacher self-efficacy beliefs (including outcome expectancies). The linkage between the left side of the model shown in Figure 1 suggests that teacher change processes (e.g., stages of concern in the CBAM model) result from teacher professional development opportunities and strengthened teacher self-efficacy beliefs. The combined and direct influence of teacher professional development, teacher self-efficacy beliefs, and teacher change *processes* is shown by the unidirectional arrow linking these variables to enhanced teaching and learning practices.

The goal of enhanced teaching and learning practices, as shown in the model, is to increase student learning and achievement. Though not shown in Figure 1, implementation of new teaching and learning practices that enhance student outcomes would be expected over time to strengthen teacher self-efficacy beliefs. The conceptual model shown in Figure 1 also assumes that interactions and linkages between variables can lead to negative relationships and outcomes. Poor teacher professional development efforts, for example, would do little to strengthen teachers' self-efficacy beliefs. Similarly, weak teacher self-efficacy beliefs would do little to move teachers through teacher change *processes* necessary to enhance the quality of teaching and learning and subsequent student learning and achievement.

A key assumption of the conceptual framework shown in Figure 1 is *context dependency*. Thus, it would be expected that linkages between variables in the framework would be influenced by the history of teacher professional development experiences (both positive and negative), years of teaching experience, pedagogical content knowledge, content knowledge, grade level taught, attitudes toward student learning, student socioeconomic status, and a host of other factors. A case can be made that each of these variables is linked to the strength of self-efficacy beliefs and engagement in change *processes*. In science education, for instance, there has been much discussion about educational reform (e.g., American Association for the Advancement of Science, 1989, 1993) and teaching practices such as student-centered and/or inquiry-based teaching and learning (National Research Council, 1996, 2000, 2012). The conceptual framework shown in Figure 1 shows the interactive influences of the quality of teacher professional development, self-efficacy beliefs (including efficacy outcome expectations), teacher change *processes*, and enhanced teaching and learning practices. The interactions and influences among these variables would be expected to be quite different in teacher professional development programs for experienced teacher leaders as opposed to professional development programs for novice teachers. Novice teachers would be expected to have weaker self-efficacy beliefs and outcome expectancies about student-centered teaching and learning practices than highly experienced teachers. Also, these two teacher groups would be expected to progress through the CBAM SoC at quite different rates, with novice teachers needing more time to work through their personal concerns than highly experienced teachers. The task dimension of the SoC in which beginning science teachers are focused on core task concerns such as time, management, efficiency, and constraints would be expected to differ from highly experienced science teachers considering adopting inquiry-based science teaching and learning practices.

The components of the conceptual framework, the assumptions on which it is based and findings from literature reviewed suggest several lines of inquiry that might be pursued in future research. Our review of the literature identified some studies linking teacher professional development and teacher self-efficacy beliefs (e.g., Khourey-Bowers & Simonis, 2004; Lakshmanan et al., 2011; Martin et al., 2008; Posnanski, 2002; Powell-Moman & Brown-Schild, 2011; Roberts et al., 2001; Ross & Bruce, 2007; Sinclair et al., 2011) and a call for including self-efficacy as a consideration in designing teacher professional development programs (Ballone-Hartzell & Czerniak, 2001; Bray-Clark & Bates, 2003). There have been few studies of teacher professional development and teacher change (e.g., Boyle et al., 2005; Guskey, 2002; Smith & Gillespie, 2007), and fewer on teacher change *processes* (Greensfeld & Elkad-Lehman, 2007) and a few studies of teacher self-efficacy and teacher change *processes* (e.g., Ellett et al., 2012; McKinney et al., 1999). Given our conceptual framework, what seems needed in future research are studies of all three of these important variables (teacher self-efficacy beliefs, teacher change *processes*, and teacher professional development,) and importantly, linkages among these variables and teacher and student outcomes. The section that follows includes a brief discussion of core findings from our review of literature, explicating our conceptual framework (Figure 1), and future research.

DISCUSSION, CONCLUSION, AND FUTURE RESEARCH POSSIBILITIES

This chapter was designed to conceptually integrate three frameworks: teacher self-efficacy beliefs, teacher change processes and teacher professional development, and. A conceptual framework (Figure 1) was developed from our understanding of a synthesis of the research literature. In explaining components of the conceptual framework a series of research questions was generated to examine linkages among the variables comprising the conceptual framework. By way of summary, there are several conclusions derived from our work that follow.

There is an apparent disconnect in the research literature among the variables constituting the conceptual framework. To better understand positive change in teacher practices and student outcomes the linkage among these variables needs to be explicated in greater detail. Teacher change is a rather complex phenomenon that necessitates more complex conceptual frameworks for comprehending the effectiveness of teacher professional development programs and how effects are mediated by important intervening variables such as teacher self-efficacy beliefs and change processes. We believe that research involving the variables and their inter-relationships shown in Figure 1, though not all-inclusive, are steps toward achieving this goal.

The framework shown in Figure 1 suggests a number of important research questions that might be addressed in future research. For example, What is the relationship between teacher change *processes* and teacher self-efficacy and outcome expectancy beliefs? What linkages can be established between teacher professional development programs and teacher change *processes*? Which variables in the conceptual framework (teacher professional development, teacher self-efficacy and outcome expectancy beliefs, and teacher change processes) are most important in enhancing (positively changing) teaching and learning practices? What is the role of teacher self-efficacy and outcome expectancy beliefs and teacher change *processes* in sustaining change in teaching and learning practices over time? How much of the variation in student achievement (particularly in student centered practices) can be predicted, accounted for, or explained by the combination of variables in the conceptual framework presented in Figure 1? What characteristics of teacher professional development contribute the most to teacher change *processes* through the mediating role of teacher self-efficacy and outcome expectancy beliefs? Each of these questions, of course, can be explored in relationship to a variety of context variables such as those described above.

REFERENCES

American Association for the Advancement of Science. (1989). *Science for all\ Americans.* Washington, DC: Author.
American Association for the Advancement of Science. (1993). *Benchmarks for science literacy.* New York, NY: Oxford University Press.
Andersen, A. M., Dragsted, S., Evans, R. H., & Sorensen, H. (2004). The relationship between changes in teachers' self-efficacy beliefs and science teaching environment of Danish first-year elementary teachers. *Journal of Science Teacher Education, 15,* 25–38.
Ballone-Hartzell, L., & Czerniak, C. M. (2001). Teachers' beliefs about accommodating students' learning styles in science classes. *Electronic Journal of Science Education, 6*(2).
Bandura, A. (1997). *Self-efficacy: The exercise of control.* New York, NY: W.H. Freeman.
Borko, H. (2004). Professional Development and Teacher Learning: Mapping the Terrain. *Educational Researcher, 33*(8), 3–15.
Boyle, B., Lamprianou, I., & Boyle, T. (2005). A longitudinal study of teacher change: What makes professional development effective? Report of the second year of the study. *School Effectiveness and School Improvement, 16*(1), 1–27.
Bray-Clark, N., & Bates, R. (2003). Self-Efficacy beliefs and teacher effectiveness: Implications for professional development. *Professional Educator, 26*(1), 13–22.
Brownell, M. T., & Pajares, F. (1999). Teacher efficacy and perceived success in mainstreaming students with learning and behavior problems. *Teacher Education and Special Education, 22,* 154–164.
Chester, M. D., & Beaudin, B. Q. (1996). Efficacy beliefs of newly hired teachers in urban schools. *American Educational Research Journal, 33,* 233–257.
Christou, C., Eliophotou-Menon, M., & Philippou, G. (2004). Teachers' concerns regarding the adoption of a new mathematics curriculum: An application of CBAM. *Educational Studies in Mathematics, 57,* 157–176.

Clarke, D., & Hollingsworth, H. (2002). Elaborating a model of teacher professional growth. *Teaching and Teacher Education, 18*, 947–967.

Davis, E. A., Petish, D., & Smithey, J. (2006). Challenges new science teachers face. *Review of Educational Research, 76*, 607–652.

Driksen, D. J., & Tharp, M. D. (1997). Utilizing the concern-based adaption model to facilitate systemic change. *Technology and Teacher Education Annual,* 1064–1067.

Ellett, C. D., Demir, A., Martin-Hansen, L., Awong-Taylor, J., & Vandergrift, N. (2012, April). *Self-efficacy, organizational culture, and change process correlates of faculty engagement in a new policy-based initiative to improve teaching and learning.* Paper presented at the annual meeting of the American Educational Research Association, Vancouver, BC.

Evans, R. H. (2011, September). *Active strategies during inquiry-based science teacher education to improve long-term teacher self-efficacy.* Paper presented at the annual meeting of the European Science Education Research Association. Lyon, France.

Fitchett, P. G., Starker, T. V., & Salyers, B. (2012). Examining culturally responsive teaching self-efficacy in a preservice social studies education course. *Urban Education, 47*, 585–611.

Fullan, M. (1992). *Successful school improvement.* Philadelphia: Open University Press.

Fullan, M. (1999). *Change forces: The sequel.* Falmer, London.

Fullan, M. (2007). *The new meaning of educational change.* New York, NY: Teachers College Press.

Gado, I., Ferguson, R., & van 't Hooft (2006). Inquiry-based instruction through handheld-based science activities: Preservice teachers' attitude and self-efficacy. *Journal of Technology and Teacher Education, 14*, 501–529.

Greensfeld, H., & Elkad-Lehman, I. (2007). An analysis of the processes of change in two science teachers educators' thinking. *Journal of Research in Science Teaching, 44*, 1219–1245.

Guskey, T. R. (2002). Professional development and teacher change. *Teachers and Teaching: Theory and Practice, 8*, 381–391.

Hall, G. E., & Hord, S. (1987). *Change in schools: Facilitating the process.* Albany, NY: State University of New York Press.

Hall, G. E., & Hord, S. M. (2001). *Implementing change: Patterns, principles, and potholes.* Boston, MA: Allyn & Bacon.

Huberman, M. (1983). Recipes for busy teachers. *Knowledge: Creation, Diffusion, Utilization, 4*, 478–510.

Huberman, M. (1995). Professional careers and professional development: Some intersections. In Guskey, T., & M. Huberman (Eds.), *Professional development in education: New paradigms and practices* (pp. 193–224). New York, NY: Teachers College Press.

James, R. K., & Hall, G. (1981). A study of the concerns of science teachers regarding an implementation of ISCS. *Journal of Research in Science Teaching, 18*, 479–487.

Khourey-Bowers, C., & Simonis, D. G. (2004). Longitudinal study of middle grades chemistry professional development: Enhancement of personal science teaching self-efficacy and outcome expectancy. *Journal of Science Teacher Education, 15*, 175–195.

Lakshmanan, A., Heath, B. P., Perlmutter, A., & Elder, M. (2011). The impact of science content and professional learning communities on science teaching efficacy and standards-based instruction. *Journal of Research in Science Teaching, 48*, 534–551.

Loucks-Horsley, S., Love, N., Stiles, K., Mundry, S., & Hewson, P. W. (2003). *Designing professional development for teacher of science and mathematics* (2nd ed.). Thousand Oaks, CA: Corwin Press, Inc.

Lumpe, A. T., Haney, J. J., & Czerniak, C.M. (2000). Assessing teachers' beliefs about their science teaching context. *Journal of Research in Science Teaching, 37*, 275–292.

Malone, M. (1984). Concerns based adoption model (CBAM): Basis for an elementary science methods course. *Journal of Research in Science Teaching, 21*, 755–768.

Martin, J. J., McCaughtry, N., McCaughtry, P., & Cothran, D. (2008). The influences of professional development on teachers' self-efficacy toward educational change. *Physical Education and Sport Pedagogy, 13*, 171–190.

McKinney, M., Sexton, T., & Meyerson, M. J. (1999). Validating the efficacy-based change model. *Teaching and Teacher Education, 15*, 471–485.

Morris, L. V. (2004). Self-efficacy in academe: Connecting the belief and the reality. *Innovative Higher Education, 28*(3), 159–162.

189

National Research Council. (1996). *National science education standards. National committee for science education standards and assessment.* Washington, DC: National Academy Press.

National Research Council (2000). *Inquiry and the national science education standards. A guide for teaching and learning.* Washington, DC: National Academy Press.

National Research Council. (2012). *A framework for K-12 science education: Practices, crosscutting concepts, and core ideas.* Washington, DC: National Academies Press.

Pajares, F. (1997). Current directions in self-efficacy research. In M. Maehr & P. R. Pintrich (Eds.), *Advance in motivation and achievement,* (Vol. 10, pp. 1–49). Greenwich, CT: JAI Press.

Posnanski, T. J. (2002). Professional development programs for elementary science teachers: An analysis of teacher self-efficacy beliefs and a professional development model. *Journal of Science Teacher Education, 13,* 189–220.

Powell-Moman, A. D., & Brown-Schild, V. B. (2011). The influence of a two-year professional development institute on teacher self-efficacy and use of inquiry-based instruction. *Science Educator, 20,* 47–53.

Richardson, V. (1990). Significant and worthwhile change in teaching practice. *Educational Researcher, 13*(7), 10–18.

Richardson, V., & Placier, P. (2001). Teacher change. In V. Richardson (Ed.), *Handbook of research on teaching* (4th ed., pp. 905–947). Washington, DC: American Educational Research Association.

Roberts, J. K., Henson, R. K., Tharp, B. Z., & Moreno, N. P. (2001). An examination of change in teacher self-efficacy beliefs in science education based on the duration of inservice activities. *Journal of Science Teacher Education, 12,* 199–213.

Ross, J., & Bruce, C. (2007). Professional development effects on teacher efficacy: Results of randomized field trial. *Journal of Educational Research, 101*(1), 50–60.

Senger, S. E. (1999). Reflective reform in mathematics: The recursive nature of teacher change. *Educational Studies in Mathematics, 37,* 199–221.

Sinclair, B. B., Naizer, G., & Ledbetter, C. (2011). Observed implementation of a science professional development program for K-8 classrooms. *Journal of Science Teacher Education, 22,* 579–594.

Smolleck, L. A., & Mongan, A. M. (2011). Changes in preservice teachers' self-efficacy: From science methods to student teaching. *Journal of Educational and Developmental Psychology, 1*(1), 133–145.

Smith, C., Hofer, J., Gillespie, M., Solomon, M., & Rowe, K. (2003). *How teachers change: A study of professional development in adult education. NCSALL reports #25.* Boston, MA: National Center for the Study of Adult Learning and Literacy.

Smith, C., & Gillespie, M. (2007). Research on professional development and teacher change: Implications for adult basic education. In Comings, J., Garner, B., & Smith, C. (Eds.), *Review of adult learning and literacy: Connecting research, policy, and practice* (pp. 205–244). Mahwah, NJ: Lawrence Erlbaum.

Woolfolk Hoy, A., & Burke-Spero, R. (2005). Changes in teacher efficacy during the early years of teaching: A Comparison of four measures. *Teaching and Teacher Education, 21,* 343–356.

AFFILIATIONS

Kadir Demir
Department of Middle and Secondary Education
College of Education
Georgia State University

Chad D. Ellett
CDE Research Associates, Inc.
Watkinsville, GA

ÁNGEL VÁZQUEZ-ALONSO & MARÍA-ANTONIA
MANASSERO-MAS

SCIENCE TEACHERS' BELIEFS ABOUT NATURE OF SCIENCE AND SCIENCE-TECHNOLOGY-SOCIETY ISSUES: CROSS-CULTURAL RESULTS THROUGH A NEW STANDARDIZED ASSESSMENT

AUTHOR'S NOTE

Research project edu2010-16553 funded by a grant of the r + d + i national program of the Ministry of Science and Innovation (Spain).

Correspondence to: Ángel Vázquez-Alonso, Departamento de pedagogía aplicada y psicología de la educación. University of the Balearic Islands, Edificio Guillem Cifre de Colonya, Carretera de Valldemossa, km. 7.5 07122 - Palma de Mallorca, Spain. Email: angel.vazquez@uib.es, ma.manassero@uib.es

SCIENCE TEACHERS' BELIEFS ABOUT NATURE OF SCIENCE AND
SCIENCE-TECHNOLOGY-SOCIETY ISSUES: CROSS-CULTURAL
RESULTS THROUGH A NEW STANDARDIZED ASSESSMENT

Teachers' educational beliefs are important because beliefs influence and guide teaching behavior and decisions within classrooms. In the case of science education, the role of beliefs is much more demanding, as science consists of a vast array of knowledge components (facts, laws and theories), but also an innovative and interdisciplinary set of historical, social and epistemological components that deeply permeate the former (contents "about" science). The former is oriented towards knowing science. The latter, however, means an understanding and awareness of science as a way of knowing. The knowledge about science view is for the judgment and critical appraisal of science and is influenced by beliefs. Components of science that correspond to features of history, philosophy and sociology of science, and the intersection of these areas, has been labeled in the literature as ideas on science, views about science, technology and society (STS), nature of science (NOS) - herein forward STS-NOS (Osborne, Collins, Ratcliffe, Millar & Duschl, 2003; Tsai, 2007).

The components about science are part of the curricula in many countries, and a teacher must be competent in order to implement them in the classroom (Aikenhead & Ryan, 1992; Lederman, 1992; Osborne et al., 2003). However, the beliefs of a teacher influence the enactment of science in a classroom. By assessing and understanding the beliefs of teachers it is possible to support teachers as they enact curricula.

R. Evans et al., (Eds.), The Role of Science Teachers' Beliefs in International Classrooms, 191–206.

This chapter is about the assessing teachers' beliefs in order to enhance their instruction. Specifically, it presents the process of diagnosing teachers' beliefs about STS-NOS issues through a paper and pencil tool, which uses a new approach. The item pool and method that is proposed in this chapter can be a useful resource for teachers, whether as a flexible, standardized instrument to assess students' beliefs or as a guide for science curriculum development on STS-NOS issues (Vazquez-Alonso & Manassero-Mas, 1999; Vázquez-Alonso, Manassero-Mas & Acevedo-Díaz, 2006).

CONCEPTUAL ISSUES ABOUT BELIEFS

Like other general constructs in education, beliefs are difficult to define. This difficulty is evident in research areas outside of psychology, such as science education (Pajares, 1992). To overcome the problem of a definition, Shrigley and Koballa (1992) suggested that beliefs could be understood through social psychology.

When looking at social psychology, the concept of belief is closely related to attitude. An attitude is a person's global psychological tendency towards an object that is expressed by evaluating the object with some degree of favor or disfavor; a belief is a specific disposition towards one attribute of the object. Eagly and Chaiken (1993) offer this perspective when they say:

"The assumption is common among attitude theorists that people have beliefs about attitude objects and that these beliefs are in some sense the basic building blocks of attitudes (p. 103)."

Thus, an attitude is made of and represents a synthesis of several beliefs the person construes about specific attributes of an object.

Beliefs are essentially judgmental and evaluative claims. According to the three-component response model of attitude, an attitude triggers three evaluative responses: cognitive, affective, and behavioral. The cognitive class refers to the thoughts (knowledge) people hold about an attitude object. The affective responses are evaluative judgments about the object, which are conceptualized as beliefs (Pajares, 1992).

Depending on the object, attitudes and beliefs are referred to in a variety of ways, which can consist of cognitions, knowledge, opinions, views, information, and inferences.

Beliefs arise in the persons' minds as a fruit of the quality and intensity of the information, experiences, and inferences about an object. However, not all beliefs are equally strong and accessible at any moment. When most beliefs represent favorable/unfavorable attributes on the object, a positive/negative attitude is formed. The most accessible and the strongest beliefs are the most influential on the person's attitude and subsequent behavior (Manstead & Hewstone, 1996).

Teachers' beliefs maintain a continuous and mutual interaction with knowledge, which has knowledge influencing beliefs, and beliefs influencing knowledge. This ongoing interaction may not be equivalent between beliefs and knowledge. In fact there may be more of an emphasis on beliefs. Specifically, for teachers, their beliefs on teaching and learning are primarily inferred from their personal observations and experiences, and less from professional books, journals, training courses or conferences (Grossman, 1995).

As teachers are significantly influenced by beliefs, they are subject to evaluative inconsistency. This is when a person can simultaneously hold positive and negative beliefs toward the same object, thus harboring inconsistent attitudes, emotions or behaviors. The ambivalence increases when the inconsistent attributes are polarized, and when the positive and negative attributes are equally evaluated or relatively uncorrelated (Eagly & Chaiken, 1993). For teachers, this state of inconsistency can impact their instruction.

In summary, by assuming a relationship between belief and attitude, and that beliefs are emphasized in different ways, it is possible to understand an individual's position in a comprehensive manner. As STS-NOS related instruction is tied to the beliefs that teachers hold, its enactment may not always be as expected.

ASSESSMENT OF TEACHERS' BELIEFS ON STS-NOS ISSUES

STS-NOS issues refers to science as a way of knowing; that is, how science acts, builds and validates its knowledge and its relationships with society and technology. The importance of STS-NOS issues stems from their acknowledgment as a crucial component of scientific literacy (DeBoer, 2000; Hodson, 2008).

The complexity, dynamism, diversity and contentiousness of STS-NOS issues involves meta-thinking on some objects of science. To list a few: The validation of knowledge, the methods used in science, and the internal and external interactions that comprise science. Meta-thinking encompasses some knowledge about the object, but mainly focuses on beliefs and attitudes towards the object. This thinking process extends one's knowledge beyond just the object, and allows for an understanding of global approaches to science such as humanistic science (Aikenhead, 2006), whole-science (Allchin, 2011), or features of science (Matthews, 2012).

Many empirical studies using different instruments and methods have repeatedly and consistently reported on the mixed nature of teachers' beliefs on STS-NOS issues. These findings are across countries and span different years of teacher experience. From the findings, it has been generally accepted that most teachers display eclectic or mixed beliefs that do not consistently fit the current scholarly STS-NOS knowledge profile (see the revisions of Deng, Chen, Tsai & Chai, 2011; García-Carmona, Vázquez & Manassero, 2011; Lederman, 1992, 2007).

Lederman (2007) reviewed the history of the STS-NOS assessments, which spanned the evolution of educational research from quantitative to qualitative methods. He ended his review by discussing his own instrument – the Views of Nature of Science Questionnaire (VNOS). The VNOS is an open-ended questionnaire that asks students for their opinions about several STS-NOS aspects, namely empirical, tentative, inferential, creative, theory-laden, social and cultural embeddedness, scientific method, and the nature of scientific theories and laws. A typical item in the VNOS will ask a question that requires a written response. For example, the item on tentativeness of science of VNOS-B asks:

"After scientists have developed a theory (e.g. atomic theory), does the theory ever change? If you believe that theories do change, explain why we bother to teach scientific theories. Defend your answer with examples."

Over the years, different forms of the VNOS (A, B, C, D, and E) have been created and extensively used by Lederman and his colleagues in many research studies (Lederman, 2007). These forms have several of these categories, but just different examples.

In using the VNOS, it is recommended that participants be given sufficient time and that individual follow-up interviews are conducted to assure validity. The VNOS answers are often examined through a qualitative analysis. This analysis format can provide information about how a person understands or reasons about science. Rubrics have also been created to assess VNOS answers. These are often scored across informed/transitional/naïve categories, and involve statistical analysis.

However, some challenges to the VNOS are evident: qualitative analysis requires time, resources and expertise on VNOS; respondents with poor writing skills or poor NOS knowledge tend to respond with short answers that hide their beliefs and create a challenging analysis process. Furthermore, the authors suggest not using the VNOS as summative assessments, though many studies use VNOS in a summative way to test the efficacy of teaching. While the VNOS yields valid and meaningful research outcomes, it is impractical for large-scale assessments, it is difficult to collect enough data that allows comparison across researchers, and it would be difficult for teachers without NOS expertise to use it in their classrooms.

To cope with these challenges, quantitative assessments have been developed over the past decade. These assessments improve validity and minimize the chance of respondents' misinterpretation of the tools.

Previous research suggested that empirically derived tools would significantly reduce the ambiguity of language, and improve standardized and quick data collection (Aikenhead, Fleming & Ryan, 1987). For instance, Tsai and Liu (2005) developed a 5-point Likert instrument that assessed high school students' epistemological views of science (SEVs) along five subscales (social negotiation in science, invented and creative science, theory-laden science, cultural impact on science, and changing science). Also, Chen (2006), starting from some selected Views on STS (VOSTS) items, developed the 5-point Likert Views on Science and Education Questionnaire (VOSE). This was done by modifying some statements and taking into account her empirical data to revise other items.

Drawing on what has been learned with the VNOS and the VOSTS, and given our desire to monitor teachers in the area of STS-NOS, we have developed a response and an assessment method. Our assessment attempts to uncover the beliefs and attitudes of teachers in the area of STS-NOS. Furthermore, it draws on the benefits and avoids the hindrances of both quantitative and qualitative approaches.

A NEW INSTRUMENT AND METHODOLOGY TO ASSESS STS-NOS BELIEFS

The "Opinions about Science, Technology and Society Questionnaire" (Spanish acronym COCTS) is a 100-item pool that is a faithful translation and adaptation into the Spanish language and cultural context (Vázquez, Manassero & Acevedo, 2005; Vázquez-Alonso, Manassero-Mas & Acevedo-Díaz, 2006) of two questionnaires: VOSTS (Aikenhead & Ryan, 1992) and Teachers Beliefs About STS (TBA-STS) (Rubba & Harkness, 1993; Rubba, Schoneweg-Bradford & Harkness, 1996). Both questionnaires were empirically developed from interviews on students and teachers. Lederman, Wade and Bell (1998), in their analysis of empirical assessment tools, consider the VOSTS to be a valid and reliable instrument for investigating positions on the nature of science.

In the COCTS, all items use a common, simple, non-technical language, and a multiple-choice format, though the number of choices is variable. The stem (or opening statement) sets up a concrete context for the different potential answers. Each potential answer develops a particular reason that explains a specific position (belief) on the issue (Manassero, Vázquez & Acevedo, 2003).

Table 1 is an example from the COCTS item pool. In this example, the question is numbered 40211 (label F2_40211_Social decisions), and is about decision-making on socio-scientific issues. The text of the stem and the multiple choice sentences are in the center, the sentence labels are to the left, and the category assigned to each sentence by experts is on the right. The multiple choice sentences are labeled A, B, C, or D, and these are located in the sentence label, along with an A, P, or I that indicates the sentence category (appropriate, plausible or ingenuous/naïve).

The following are important attributes of this methodological approach and response model, which assesses STS-NOS beliefs through the COCTS pool:

• Respondents are not compelled to make a forced choice for one multiple-choice sentence (Vázquez-Alonso & Manassero-Mas, 1999)..
• The scaling of COCTS statements into one out of three categories (Appropriate-Adequate, Plausible, or Naïve-Ingenuous) was completed by a panel of expert judges in the areas of the history, philosophy and sociology of science and technology (Rubba & Harkness, 1993; Tedman & Keeves, 2001; Vázquez et al., 2005).
  – Adequate-Appropriate (A): the statement expresses an adequate belief;
  – Plausible (P): though not totally adequate, the statement expresses some acceptable aspects; and
  – Ingenuous-Naïve (I): the statement expresses a belief that is neither adequate nor plausible.
• A new multiple response model (MRM) that maximizes the information contained within the sentences; the respondent rates all statements along a 9-point Likert scale.

Table 1. Example item from COCTS item pool

Sentence Label	Item Text	Category
	40211 Scientists and engineers should be the ones to decide what types of energy our country will use in the future (for example, nuclear, hydro, solar, or coal burning) because scientists and engineers are the people who know the facts best.	
	Scientists and engineers should decide:	
F2_C_40211A_I_ Social decisions	A. Because they have the training and facts which give them a better understanding of the issue.	Naïve
F2_C_40211B_I_ Social decisions	B. Because they have the knowledge and can make better decisions than government bureaucrats or private companies, both of whom have vested interests.	Naïve
F2__40211C_P_ Social decisions	C. Because they have the training and facts which give them a better understanding; BUT the public should be involved, either informed or consulted.	Plausible
F2_C_40211D_A_ Social decisions	D. The decision should be **made equally**; viewpoints of scientists and engineers, other specialists, and the informed public should all be considered in decisions which affect our society.	Adequate
F2__40211E_P_ Social decisions	E. The **government** should decide because the issue is basically a political one; BUT scientists and engineers should give advice.	Plausible
F2__40211F_A_ Social decisions	F. The **public** should decide because the decision affects everyone; BUT scientists and engineers should give advice.	Adequate
F2__40211G_P_ Social decisions	G. The **public** should decide because the public serves as a check on the scientists and engineers. Scientists and engineers have idealistic and narrow views on the issue and thus pay little attention to consequences.	Plausible
F2__40211H_P_ Social decisions	H. It depends of the type of decision; it is not the same thing to decide on the nuclear disarmament or on a baby. In some cases the scientists could make the decision, but in other, the citizens or the stakeholders should make it.	Plausible

Note. The coding *C* before the tag number item means that the sentence represents an idea on which most (2/3) of the expert judges strongly agreed on its assigned category.

- The construction of a quantitative metric that produces a normalized index [-1, +1] for each statement, which represents the respondent's belief through the degree of match between belief and the experts' current conceptions; the index meaning is standardized and invariant, as it is independent of the qualities of the original sentence (Manassero, Vázquez & Acevedo, 2001).
- The statement indices can be used to further computations. The average of the item sentence indices produces a global attitudinal item index (Manassero et al., 2003), according to the idea that an attitude synthesizes several beliefs (Eagly & Chaiken, 1993).
- The indices allow the application of correlational and inferential statistics for hypothesis testing, group comparison, or to establish cut-off points for achievement levels (Vázquez-Alonso et al., 2006).

RESULT

The results displayed below correspond to the application of the new method and scoring system with 1,631 science in-service teachers of 8 different cities, within 7 Latin-speaking countries from elementary, middle, or high school teachers. Each teacher answered one randomly assigned booklet (15 items) out of two different booklets, called Form 1 (F1 n = 916) and Form 2 (F2 n = 715), which encompassed 99 and 101 statements. The booklets were designed in order to get balance between avoiding respondents' fatigue and acceptable coverage of all COCTS dimensions.

Table 2 shows the analysis of the indices obtained for the whole set of variables (questions and sentences), and contains the details about teachers' general attitudes (questions) and specific beliefs (sentences). In the table, each item is represented by a five-digit key number, where the figure corresponds successively to dimension, sub-dimensions, themes and sub-themes, which are also described by a short label behind the key number.

Table 2. Dimensions, labels and Cronbach's alpha of the assessment items included in the two questionnaire forms (F1 and F2).

STS-NOS Dimensions	Form 1 (F1) Sub-themes Items	Cronbach's alpha*	Form 2 (F2) Sub-themes Items	Cronbach's alpha*
a) Definition of science and technology	F1_10111 science	0.224	F2_10211 technology	0.839
	F1_10411 interdependence	0.322	F2_10421 interdependence quality of life	0.882
b) Science–Technology–Society interactions	F1_30111 STS interactions	0.771		

(*Continued*)

Table 2. Dimensions, labels and Cronbach's alpha of the assessment items included in the two questionnaire forms (F1 and F2). (Continued)

STS-NOS Dimensions	Form 1 (F1) Sub-themes Items	Cronbach's alpha*	Form 2 (F2) Sub-themes Items	Cronbach's alpha*
b.1) Influence of society on S&T	F1_20141 country's government politics	0.289	F2_20211 industry	0.867
	F1_20411 ethics	0.944	F2_20511 educational institutions	0.889
b.2) Influence of S&T on society	F1_40161 social responsibility for pollution	0.448	F2_40131 social responsibility information	0.787
	F1_40221 moral decisions	0.702	F2_40211 social decisions	0.911
	F1_40531 social well-being	0.727	F2_40421 application to daily life	0.883
			F2_50111 union two cultures	0.846
b.3) Internal sociology of S&T	F1_60111 motivations	0.889	F2_60521 gender equality	0.891
	F1_60611 women's under-representation	0.495	F2_70211 scientific decisions	0.887
	F1_70231 consensus decisions	0.752	F2_70711 national influences	0.863
	F1_80131 advantages for society	0.750		
c) Epistemology	F1_90211 scientific models	0.916	F2_90111 observations	0.864
	F1_90411 tentativeness	0.478	F2_90311 classification schemes	0.888
	F1_90621 scientific method	0.852	F2_90521 role of assumptions	0.827
			F2_91011 epistemological status	0.799

Cronbach's alphas are computed for the sub-sample of Spanish science teachers (n = 774).

Item Indices Across Places: General Attitudes

The indicators of the teachers' general attitude toward the STS-NOS issues posed in the 30 questions are the questions' average global indices (Figures 1 and 2). The practitioner science teachers' profiles display two striking and apparently contradictory patterns: on the one hand, a quasi-parallel change across items, and on the other hand, the relevant differences between places in many items.

The most striking feature is the fairly parallel profile across items in both F1 and F2 booklets. The points tend to be located on the same items across the different places. For instance, the items F1_30111, F1_40161, F1_60611, F2_20511, F2_40131, F2_50111, F2_60521 appear as relative maximum; the relative minimum tend to be on items F1 20411, F1_40531, F2_10211, F2_40421, F2_70211, F2_90521, and the medium indices on F1_10111, F1_70231, F2_40211, F2_70711. The quasi-parallel change across places suggests a kind of attitudinal homogeneity of the issues.

On the other hand, science teachers' overall attitudes show some relevant differences between places on many items. For most items, the effect size of the differences between the highest and lowest place is higher than 0.20 score (threshold of relevant differences according to effect size statistics). A few exceptions to these large differences between places correspond to items F1_60111, F1_80131, F2_10211, F2_20511, F2_40421, F2_60521 and F2_70711.

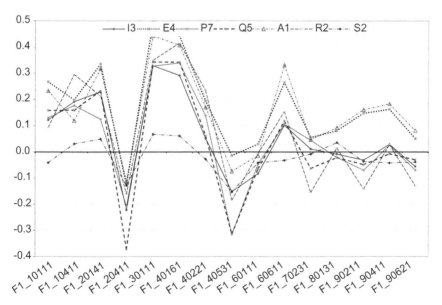

Figure 1. Mean indices for each of the 15 items of the F1 questions for the sample of practitioner science teachers across different places.

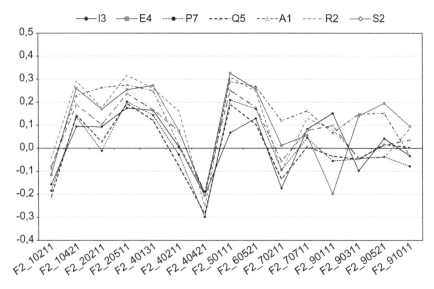

Figure 2. Mean indices for each of the 15 items of the F2 questions for the sample of practitioner science teachers across different places.

Despite this complexity of profiles, the figures show that some sites tend to be placed in the highest positions (A1, R2, M4, E4), while some other places exhibit the lowest indices (S2, P7, Q5). Further, this trend about higher and lower places holds for both booklets, thus, reinforcing the soundness of the pattern.

The item mean indices also highlight the strengths and weaknesses of teachers' understanding on STS-NOS issues through the highest (positive) or lowest (negative) indices. These issues can be analyzed by applying the effect size criterion for relevant top and bottom differences (average indices beyond one third of a mean standard deviation over/under zero score, approximately 0.1 points), which correspond to the strengths (highest positive indices) and the weaknesses (lowest negative) of questions. Table 3 shows these results.

TEACHERS' BELIEFS: SENTENCE INDICES

Each sentence conveys an attribute of the question issue; thus, representing a belief on the issue. The sentence index represents the quality of teacher's specific belief, and the analysis of the sentence indices reveals the strength of the teachers' beliefs. The overall average indices for the 200 sentences that teachers rated on 30 issues show that under one third attained relevant positive scores (d > .30 SD over zero score). Most corresponded to adequate sentences and a few corresponded to naïve sentences. Plausible sentences were absent.

Table 3. Practitioner science teachers' strengths and weaknesses drawn from the 30 questions.

Strengths	Weaknesses
F1_40161 Social responsibility contamination	F2_70211 Scientific decisions
F1_30111 STS Interaction	F2_10211 Technology
F2_50111 Union two cultures	F1_20411 Ethics
F2_20511 Educational institutions	F2_40421 Application to daily life
F2_60521 Gender equity	F1_40531 Life welfare
F1_10411 Interdependence	
F2_40131 Social responsibility information	
F2_10421 Interdependence quality of life	
F1_10111 Science	
F1_20141 Government politics a country	
F1_40221 Moral decisions	
F1_60611 Women under representation	

A small part of these sentences exhibit mean indices that lie one SD away from the overall mean (highest and lowest), which represent the teachers' most (strongest) and worst (weakest) beliefs. The statements in Table 4 are ordered by form and by decreasing value of the mean index. Regarding strengths, it is worth noting the presence of three statements corresponding to the same item, F1_40161 social responsibility for pollution. The respondents had the highest indices on the naïve statement A ("Heavy industry should be moved to underdeveloped countries to save our country and its future generations from pollution"), and two appropriate statements D and F, which were preceded by the entry "Heavy industry should NOT be moved to underdeveloped countries" as most informed beliefs (reasons) to justify the environmental-friendly decision to not move heavy industry to underdeveloped countries. These items are elaborated on in the Appendix.

Regarding weaknesses, three plausible sentences refer to gender equity in science (F2_60521). The lowest indices reveal that the majority of teachers fail to realize that the statements were not totally adequate, even though they expressed some acceptable aspects. The teachers tended to express strong dis/agreement with these statements.

Two weaknesses about epistemological issues, concerning scientific method (90621A) and the epistemological status of scientific knowledge (91011B) were noteworthy. Item 90621 stated that "The best scientists are those who follow the steps of the scientific method" and the naïve position A argues that "the scientific method ensures valid, clear, logical and accurate results. Thus, most scientists will follow the steps of the scientific method". Item 91011 addressed whether scientists

Table 4. Sentences with the highest positive indices (the best informed beliefs) and the lowest negative indices (the worst informed beliefs).

Highest positive index (strengths)	Lowest negative index (weaknesses)
FORM 1	FORM 1
F1_C_60611A_N_women's under-representation	F1_C_90621A_N_scientific method
F1_C_40161A_N_social responsibility for pollution	F1__60611H_P_women's under-representation
F1_C_40161D_A_social responsibility for pollution	F1__60111G_N_motivations
F1_C_40161F_A_social responsibility for pollution	F1__70231A_N_consensus decisions
F1_C_10411B_A_interdependence	
F1_C_40161C_A_social responsibility for pollution	
FORM 2	FORM 2
F2__60521D_A_gender equality	F2__60521E_P_gender equality
F2_C_10421H_N_interdependence quality of life	F2_C_70211A_N_scientific decisions
F2_C_60521F_A_gender equality	F2__91011B_N_epistemological status
F2_C_50111E_A_union two cultures	F2__60521A_P_gender equality
F2_C_40211D_A_social decisions	F2__60521B_P_gender equality
	F2__60521C_P_gender equality

discover or invent scientific laws, hypotheses, and theories. The naïve position B argues for discovering . . . "because laws are based on experimental facts." The low indices achieved by the previous two naïve statements mean that many teachers expressed relatively strong agreement with them, when disagreement would have been expected for a well-informed conception.

Finally, these results also reveal an inconsistency of beliefs. The comparison between the highest positive and lowest negative lists reveals that some items have sentences in both lists (for instance, items F1_40161 and F1_60611). This extreme polarization of beliefs within an item means inconsistency.

CONCLUSIONS AND DISCUSSION

This paper focuses on the assessment of teachers' beliefs on STS-NOS components. The presented instrument (a scaled, open-ended, flexible, multiple-choice, paper-and-pencil pool of items), and methodology (multiple response model and scaling scores that produces standardized indices) offer a fast, easy, cheap, valid, reliable, and efficient way to assess STS-NOS beliefs. Furthermore, the tool is applicable to large samples, allows statistical hypothesis testing and correlational analysis, and, simultaneously facilitates the contrast with other tools.

The main finding of the tool is the possibility of setting up detailed, individual and group profiles of beliefs that are quantitatively-based and qualitatively-developed. These results herein showed that one third of teachers' beliefs were relevantly positive, which offers some points worthy of discussion. First, many of the highest positive beliefs align with ideas which the group of expert judges strongly agreed with. Second, teachers hold at least one belief with a relevant positive index on almost every STS-NOS issue. These positive beliefs could be used as hooks for the re-construction and the enhancement of teachers' beliefs.

Another empirical finding pinpoints the ambivalence of teachers' beliefs, as suggested by theorists (Eagly & Chaiken, 1993). This finding specifically suggests that teacher thinking seems to be much closer to superficiality than to rationality, as teachers often express agreement on two logically incompatible sentences. This new tool and method might add reliable evaluations of inconsistent beliefs; a crucial area that can improve research on teachers beliefs (Kind & Barmby, 2011).

Mainstream research on STS-NOS acknowledges the difficulties of teaching STS-NOS issues, and the need of explicit and reflective approaches (Abd-El-Khalick, 2012). Teachers' understanding on STS-NOS revealed that on one hand, one third of the teachers had informed beliefs; on the other hand, two third of beliefs were either negative or not relevantly positive as expected for a qualified science teacher. Furthermore, most beliefs were ambivalent (superficial, lacking coherence and consistence). Thus, there is support for explicit and reflective teacher training as mean to overcome inconsistency (Abd-El-Khalick, 2012; Matthews, 2012).

Summing up, this tool is a beginning to understanding the STS-NOS beliefs of teachers. With further articulation of the validity and reliability of this tool and process, this instrument may be plausible for large scale studies. At this point in time, this is a novel format that can look at the beliefs of teachers who may be teaching STS-NOS areas.

APPENDIX

Text of the sentences that represent the strongest and weakest teachers' beliefs (highest and lowest mean indices) on some selected questions.

Question Stem	Strong Beliefs
40161 Heavy industry has greatly polluted very much. Therefore, it is a responsible decision to move heavy industry to underdeveloped countries where pollution is not so widespread.	A. Heavy industry should be moved to underdeveloped countries to save our country and its future generations from pollution. C. It doesn't matter where industry is located. The effects of pollution are global.

(Continued)

Continued

Question Stem	Strong Beliefs
Heavy industry should NOT be moved to underdeveloped countries:	D. because moving industry is not a responsible way of solving pollution. We should reduce or eliminate pollution here, rather than create more problems elsewhere.
	F. because pollution should be confined as much as possible. Spreading it around would only create more damage.
60611Today in our country, there are many more male scientists than female scientists. The MAIN reason for this is:	A. males are stronger, faster, brighter, and better at concentrating on their studies.
60521 When doing science or technology, a good female scientist would carry out the job basically in the same way as a good male scientist. There is NO difference between female and male scientists in the way they do science:	D. because women and men are the same in terms of what is needed to be a good scientist. F. because any differences in the way scientists do science are due to differences between individuals. Such differences have nothing to do with being male or female.

Question Stem	Weak Beliefs
60611Today in our country, there are many more male scientists than female scientists. The MAIN reason for this is:	H. There are NO reasons for having more male scientists than female scientists. Both sexes are equally capable of being good scientists, and today the opportunities are equal.
60521 When doing science or technology, a good female scientist would carry out the job basically in the same way as a good male scientist. There is NO difference between female and male scientists in the way they do science:	A. because all good scientists carry out the job the same way. B. because female and male scientists experience the same training. C. because overall women and men are equally intelligent. E. because everyone is equal, no matter what the job.

REFERENCES

Abd-El-Khalick, F. (2012). Examining the Sources for our understandings about science: Enduring conflations and critical issues in research on nature of science in science education. *International Journal of Science Education, 34*(3), 353–374.

Aikenhead, G. S. (2006). *Science education for everyday life: Evidence-based practice.* New York, NY: Teachers College, Columbia University.

Aikenhead, G. S., & Ryan, A. G. (1992). The development of a new instrument: Views on science-technology-society (VOSTS). *Science Education, 76*(5), 477–491.

Aikenhead, G. S., Fleming, R. G., & Ryan, A. G. (1987). High school graduates' beliefs about science-technology-society. I. Methods and issues in monitoring students views. *Science Education, 71*, 145–161.

Allchin, D. (2011). Evaluating knowledge of the nature of (whole) science. *Science Education, 95*, 518–542.

Chen, S. (2006). Development of an instrument to assess views on nature of science and attitudes toward teaching science. *Science Education, 90*(5), 803–819.

DeBoer, G. E. (2000). Scientific literacy: Another look at its historical and contemporary meanings and its relationship to science education reform. *Journal of Research in Science Teaching, 37*, 582–601.

Dogan, N., & Abd-El-Khalick, F. (2008). Turkish grade 10 students' and science teachers' conceptions of nature of science: A national study. *Journal of Research in Science Teaching, 45*(10), 1083–1112.

Deng, F., Chen, D.-T., Tsai, C.-C. & Chai, C.-S. (2011). Students' views of the nature of science: A critical review of research. *Science Education, 95*, 961–999.

Eagly, A. H., & Chaiken, S. (1993). *The psychology of attitudes*. Fort Worth, TX: Harcourt Brace College Publishers.

García-Carmona, A., Vázquez, A., & Manassero, M. A. (2011). Estado actual y perspectivas de la enseñanza de la naturaleza de la ciencia: una revisión de las creencias y obstáculos del profesorado. [Current status and prospects about teaching the nature of science: A review of teachers' beliefs and obstacles] *Enseñanza de las Ciencias, 29*(3), 403–412.

Grossman, P. L. (1995). Teachers' knowledge. In L.W Anderson (ed.), *Encyclopedia of teaching and teacher education* (pp. 20–24). New York: Pergamon.

Hodson, D. (2008). *Towards scientific literacy: A teachers' guide to the history, philosophy and sociology of science*. Rotterdam: Sense Publishers.

Kind, P., & Barmby, P. (2011). Defending attitude scales. In I. M. Saleh & M. S. Khine (Eds.), *Attitude research in science education: Classic and contemporary measurements* (pp. 117–135). Charlotte, NC: Information Age Publishing Inc.

Lederman, N. G. (1992). Students' and teachers' conceptions of the nature of science: A review of research. *Journal of Research in Science Teaching, 29,* 331–359.

Lederman, N. G. (2007). Nature of science: Past, present, and future. In S. K. Abell & N. G. Lederman (Eds.), *Handbook of research on science education* (pp. 831–879). Mahwah, NJ: Lawrence Erlbaum Associates.

Lederman, N. G., Wade, P. D., & Bell, R. L. (1998). Assessing understanding of the nature of science: A historical perspective. In W. F. McComas (Ed.), *The nature of science in science education: Rationales and strategies* (pp. 331–350). Dordrecht, The Netherlands: Kluwer Academic Publishers.

Manassero, M. A., Vázquez, A., & Acevedo, J. A. (2001). *Avaluació dels temes de ciència, tecnologia i societat*. [Assessment of science, technology and society issues] Palma de Mallorca: Govern de les Illes Balears.

Manassero, M. A., Vázquez, A., & Acevedo, J. A. (2003). *Cuestionario de opiniones sobre ciencia, tecnología y sociedad (COCTS)*. [Questionnaire of opinions on science, technology and society] Princeton, NJ: Educational Testing Service.

Manstead, A. S. R., & Hewstone, M. (Eds.). (1996). *The Blackwell encyclopedia of social psychology*. Malden, MA: Blackwell Publishing.

Matthews, M. R. (2012). Changing the focus: From nature of science (NOS) to features of science (FOS). In M. S. Khine (Ed.), *Advances in nature of science research. concepts and methodologies*, (pp. 3–26), Heidelberg, Springer Dordrecht.

Osborne, J., Collins, S., Ratcliffe, M., Millar, R., & Duschl, R. (2003). What ideas-about-science should be taught in school science? A Delphi study of the expert community. *Journal of Research in Science Teaching, 40*(7), 692–720.

Pajares, M. F. (1992). Teachers' beliefs and educational research: Cleaning up a messy construct. *Review of Educational Research, 62*(3), 307–332.

Rubba, P. A., & Harkness, W. L. (1993). Examination of pre-service and in-service secondary science teachers' conceptions about science-technology-society interactions. *Science Education, 77*(4), 407–431.

205

Rubba, P. A., Schoneweg-Bradford, C., & Harkness, W. L. (1996). A new scoring procedure for the Views on Science-Technology-Society instrument. *International Journal of Science Education, 18*(4), 387–400.

Shrigley, R. L., & Koballa, T. R. (1992). A decade of attitude research based on Hovland's learning theory model. *Science Education, 76*(1), 17–42.

Tedman, D. K., & Keeves, J. P. (2001). The Development of Scales to Measure Students' Teachers' and Scientists' Views on STS. *International Education Journal, 2*, 20–48.

Tsai, C.-C. (2007). Teachers' scientific epistemological views: the coherence with instruction and students' views. *Science Education, 91*(2), 222–243.

Tsai, C.-C., & Liu, S-Y. (2005). Developing a multi-dimensional instrument for assessing students' epistemological views toward science. *International Journal of Science Education, 27*(13), 1621–1638.

Vázquez-Alonso, A. & Manassero-Mas, M. A. (1999). Response and scoring models for the 'Views on Science-Technology-Society' Instrument. *International Journal of Science Education, 21*(3), 231–247.

Vázquez, A., Manassero, M. A., & Acevedo, J. A. (2005). Quantitative analysis of complex multiple-choice items in science technology and society: Item scaling. *Revista Electrónica de Investigación Educativa, 7*(1). Retrieved from http://redie.uabc.mx/vol7no1/contents-vazquez.html

Vázquez-Alonso, A., Manassero-Mas, M. A., & Acevedo-Díaz, J. A. (2006). An Analysis of Complex Multiple-Choice Science-Technology-Society Items: Methodological Development and Preliminary Results. *Science Education, 90*(4), 681–706.

AFFILIATIONS

Ángel Vázquez-Alonso
Departament of Applied Pedagogy and Educational Psychology
Faculty of Education
University of the Balearic Islands, Spain

María-Antonia Manassero-Mas
Department of Psychology
Faculty of Psychology
University of the Balearic Islands, Spain

JAMES J. WATTERS

CHALLENGES OF ELEMENTARY SCIENCE TEACHING

An Australian Perspective

INTRODUCTION

This chapter profiles research that has explored the role of affect in the teaching of science in Australia particularly on primary or elementary science education. Affect is a complex set of characteristics that relate to the interactions between an individual's knowledge and emotional responses to a stimulus. Thus, there are many dimensions and theoretical frameworks that inform our understanding of how and why people behave in particular ways. Social cognitive theory has proven to be an effective lens to examine behaviour. It argues that human engagement in any behaviour, such as the teaching of science, is regulated by intentional cognitive processes (i.e., human agency) in response to observations of others engaged in that behaviour (Bandura, 2001). By reflecting on one's own actions, people develop beliefs about their capability to perform the actions and their control over the outcomes of those actions. As Bandura (2001) states, "Unless people believe they can produce desired results and forestall detrimental ones by their actions, they have little incentive to act or to persevere in the face of difficulties" (p. 10). Such efficacy beliefs are powerful regulators to be considered in explaining how teachers engage in the teaching of science.

The first section presents a brief historical and contextual overview to highlight the contemporary issues confronting science teaching and learning in Australian schools. A second section focuses on the early years of learning and the debates about the purpose of science in primary schools. The third and main section of the chapter examines the research related to affect. The chapter concludes with some of the major policy changes that are influencing the teaching of science in Australia.

THE AUSTRALIAN CONTEXT

Historical Background

Australia is a federation of six states and two semi-autonomous territories. The responsibility for education lies with the various state governments. Until recently, each state has developed its own curricula, but choice of specific content and pedagogical approaches remains with the individual schools or teachers to develop.

R. Evans et al., (Eds.), The Role of Science Teachers' Beliefs in International Classrooms, 207–225.
© 2014 Sense Publishers. All rights reserved.

Teachers rarely use commercial textbooks for primary science but do have access to a range of resource and support materials. Approaches to schooling and curricula across the states vary in details but essentially schooling is organised around three broad phases: non-compulsory pre-school/kindergarten, compulsory schooling from grade 1 (six years of age) to grade 10 and non-compulsory senior schooling in years 11 and 12. The division of compulsory schooling into primary and junior secondary varies across states so that years 1-6 are primary and years 7-10 secondary. In some states, year 7 is located in primary school. Early childhood, primary, and secondary teachers undertake different pre-service teacher educational programs.

The economic climate and shifts in population dynamics over the past 20 years have increased pressure on standardising the education system across the country. The first attempt to bring some consistency in curriculum was the design and release of a declaration on schooling in 1989 – the Hobart Declaration (Australian Education Council, (1989). This document endorsed by all state education ministers identified the nationally agreed goals for schooling and indicated that national collaborative curriculum development would be undertaken in eight key learning areas which included Science. In 1994, a National Statement on Science for Australian Schools (Curriculum Corporation, 1994), which identified science as a key learning area (KLA), was released. The National Statement on Science for Australian schools and an accompanying document, Science – a curriculum profile for Australian schools, provided guidelines for the future development of science teaching throughout Australia. Science – a curriculum profile for Australian schools organised the content of science into five strands: Working scientifically, energy and change, natural and processed materials, earth and beyond, and life and living. Working scientifically was a process strand emphasising inquiry skills, and the four other strands were concept strands. Subsequently, over the next decade as states revised their curricula particularly at the primary school, these themes were adopted to varying degrees to inform state syllabi. Significantly, constructivist philosophies were influential in the development and implementation of new science curricula. This document was followed by further ministerial declarations released in roughly ten-year intervals that set in place a vision for education in Australia culminating in the release of a national curriculum in 2010. All states and territories were committed to work toward implementation by 2013. Science is accorded high status with a requirement that it be taught to all students in each year of schooling from foundation (pre-school) to year 10 (Australian Curriculum Assessment Reporting Authority [ACARA], 2012). Indicative time allocations per week have been allocated to individual key learning areas.

It is in this context of the rationalisation of curriculum across the nation that attention is turned to the teaching of science in the primary years of schooling.

Status of science teaching in schools

Studies conducted in Australia on the teaching of science or student performance in science have been generally critical of teacher confidence and beliefs in the purpose of science in the curriculum. This section will address both these issues and suggest that a focus on a teacher's canonical knowledge of science is counterproductive.

Over 20 years ago a review conducted by Peter Fensham and Graeme Speedy (Department of Employment, Education, and Training [DEET], 1989) identified the low level of confidence in teaching science among primary teachers. They attributed this to the lack of science discipline knowledge and argued that pre-service teacher education programs needed to address both content and professional knowledge. However, Fensham and Northfield (1993) noted the difficulties that teacher educators faced in addressing these issues. Programs in teacher education needed to identify the appropriate knowledge and depth of knowledge. Thus, diploma level courses gave way to degree programs that included both content and professional (pedagogical) knowledge. However, in most institutions, the discipline studies were taught by science specialists separate from the educational studies staff who were often located in different faculties. The review brought into sharper focus the relationships among teacher confidence, disciplinary knowledge and the willingness to teach science, and the scope of what science was taught.

In 2000, Goodrum, Hackling and Rennie published a comprehensive analysis of the teaching of science across years 5-10 in schools in Australia. Their assessment echoed what various other stakeholders assumed:

> The actual picture of science teaching and learning is one of great variability but, on average, the picture is disappointing. Although the curriculum statements in States/Territories generally provide a framework for a science curriculum focused on developing scientific literacy and helping students progress toward achieving the stated outcomes, the actual curriculum implemented in most schools is different from the intended curriculum. In some primary schools, often science is not taught at all... Disenchantment with science is reflected in the declining numbers of students who take science subjects in the post-compulsory years of schooling. (p. viii)

Although there is considerable media criticism and concern among science educators about the status of science teaching, the reality is that by international standards Australia performs quite well. Analysis of PISA and TIMSS data generally ranks Australia in the top ten percent of participating nations and certainly at the top of many English-speaking countries (OECD, 2010). Notwithstanding these international results, assessment of scientific literacy undertaken by federal government authorities continues to fuel concern about the learning of science with findings that only between 52% and 59% of year 6 students attained an acceptable "proficiency standard" in scientific literacy based on criteria determined by the Australian Curriculum, Assessment and Reporting Authority [ACARA].

Student interest in studying science, engineering, and mathematics is at an all-time low in Australia. Numerous reports released over the past 20 years have shown that interest among Australian students to pursue science related careers is at best stagnating (Dobson, 2012; Goodrum, Druhan & Abbs, 2011). Over the same time, science and mathematics education among Australia's trading partners in the Asia Pacific region has been improving base on student performance on international tests. These two effects have contributed to a decline in the ranking of Australian students' success on international tests.

The perceived poor performance of primary students is frequently attributed to poor teaching and the lack of confidence and knowledge of primary and early years teachers (Appleton, Ginns & Watters, 2000; Appleton & Symington, 1996; Australian Academy of Technological Sciences and Engineering [ATSE], 2002; Department of Employment, Education and Training [DEET], 1989; Harlen, 1997; Osborne & Simon, 1996). The TIMSS assessment, besides probing student performance, also asked year 4 teachers whether they felt prepared to teach science and mathematics. The response in 2007 was that only 46% of these Australian primary teachers felt "very well prepared" to teach science compared with a mean of 54% across all participating countries. These concerns were not new and were aired in Australia nearly forty years ago (Symington, 1974). There has been broad consensus that primary school teachers were not knowledgeable about science, lacked confidence to teach it; hence they avoided teaching science. The inadequacy of professional development programs and pre-service education for primary science education was frequently cited (Loucks-Horsley & Matsumoto, 1999; Venville, Wallace & Louden, 1998).

The current approach to teaching science has been described by Osborne (2007) as "arcane". He asserts that an idealized view of science as objective, detached and value free is presented in schools and that science teaching has focussed on building in students an understanding of scientific facts. This form of education might suit a minority of students intrinsically interested in pursuing science but it fails to engage the majority of students. Most students will need to use science or understand aspects of science to be productive members of a democratic society. Fensham (1993) argued that teaching "science for all" was one of the major challenges facing contemporary science education. The implications is that primary teachers as non-subject specialist teachers need to have a knowledge base in science that is quite distinct from that of say a secondary teacher. Secondary teachers teach students who generally choose to learn science because of intrinsic interests or as a pathway to a science related career.

The view that science is a body of knowledge to be learnt and passed on to succeeding generation prevails among many teachers and indeed policy makers. These views need to be challenged as primary teachers need to be truly scientifically literate in the sense that they not only have a general understanding of science but that they see how science relates to other domains of knowledge and can package instruction in ways that contextualise the content for students. If teachers have inappropriate beliefs about the nature and purpose of science, they will not attempt to reform their approaches. However, teachers work within the bounds of their schools culture which is a reflection of the values and priorities generally set by staff and principals and of course the prevailing curricula emphases. If there is leadership around science teaching and there is interest in science by the principal or some other significant person then the way that teachers approach the teaching of science changes.

To summarize, if teachers as a collective group within a school believe science teaching is about inert knowledge of content and they do not possess that knowledge, then science will be a low priority. If teachers acknowledge and believe that science is more than knowledge and incorporates an understanding that science is a social enterprise, then their own personal experiences become important and as source for building science teaching self-efficacy. Evidence that a focus on the nature of science in preservice courses has potential to influence teachers' beliefs emerges in the study by McDonald (2010).

Importance of early-years science education

The importance of science in the primary years of schooling has been argued from the perspective that early exposure to science establishes positive attitudes and dispositions, develops capacity to engage as citizens in a technological world, and contributes to the development of a scientifically literate citizen who is able to understand and contribute to social issues that involve scientific knowledge (Harlen & Qualter, 2004). Nurturing young children curiosity and inspiring them to develop positive attitudes and an appreciation of the nature of science is critical. If attitudes are developed where they think science is too hard or irrelevant to their lives, it becomes difficult to change that view once formal science is introduced in secondary school.

Teachers, principals, and schools make a difference (Hattie, 2009). Teachers need to show passion and commitment as well as competence. Principals play critical roles in leading the formation of a school climate that values science and supports teachers in the teaching of science through professional learning and resourcing. With new ideas and information, teachers feel competent in the classroom facing their students. Moreover, the feedback from students will tell teachers how successful they have been in achieving their objectives. This kind of feeling of competence not only enhances teachers' motivation, but also it encourages teachers to undertake further professional development to improve their professional competence. For primary teachers, who are generalists, conceptual knowledge of science may be less important than their commitment to implement science – given adequate resources. The affective dimension becomes a critical contribution to effective teaching.

AFFECTIVE DIMENSIONS

A number of Australian studies began to emerge in the context of an enhanced national approach to science teaching and the relocation of teacher education into a university sector. Some of the early research attempted to focus on the type of knowledge that teachers need to be effective in teaching science. Drawing upon the interactive science teaching approach developed by Biddulph and Osborne (1984) in New Zealand, Hardy, Bearlin and Kirkwood (1990) attempted to address the issue of teacher engagement in primary science. They saw the basis of teacher lack of engagement as lack of confidence in teachers own conceptual knowledge. Thus, they challenged both pre-service and in-service teachers in a year-long program of workshops to not focus on their own knowledge but on their students' knowledge. The effectiveness of this model was explored by several researchers. For instance, Appleton (1992), in a mixed methods design, investigated the change in attitudes of pre-service primary and early childhood teachers (n=139) who were exposed to a compulsory integrated method and science discipline subject taught within an education context using the Hardy et al. model. He presented students with a range of questions about their level of interest in teaching science and the perceptions of their knowledge to teacher science. Significant changes (effect sizes of the order of .6 to .8) were observed in students' perceptions of their background knowledge to teach different topics in science. In contrast, changes in response to a question probing attitude to teaching science although positive were very small (effect size of 0.2). In the interviews he conducted, it appeared that given appropriate pedagogical practices in teaching science students became more positive in their perceptions about teaching science. Key aspects of the pedagogy were described as being focussed on "gender equity, constructivism, and science as a dynamic, people-oriented subject" (p. 16). In a similar study but with in-service teachers (n=40) participating in an extended professional development program in teaching science and design technology, Aubusson and Webb (1992) found that teachers believed that knowledge and confidence were important for good teaching. Interestingly, they rated student knowledge of science as relatively unimportant which the researchers interpreted as "by rating the development of science and technology knowledge by pupils as relatively unimportant, their own lack of knowledge becomes unimportant and their self-esteem as teachers is thereby protected" (p. 23).

Baker (1994) began to question the nature of knowledge that primary teachers needed arguing that "what actually constitutes the 'science content' required by teachers of science is rather more complex than that implied by a 'background in science'" (p. 32). She perceived that interactive teaching practices and the prevailing process approach to science emphasised in many curriculum documents afforded opportunities for teachers to become "facilitators" often to the point of simply documenting students' ideas about concepts but not having to redevelop these understandings. This was a similar conclusion to that of Aubusson and Webb (1992) who found that even those teachers who were committed to science education found it difficult to describe how they should interact with children in order to promote learning in science. These studies highlighted the tensions faced by a teacher who has limited content knowledge or the necessary pedagogical content knowledge to know how to exploit what content knowledge they had to align with the needs of students. A limitation of these studies was a lack of theorisation of the key constructs. Although Baker was drawing upon Shulman's (1987) ideas of teacher knowledge, the affective dimensions of attitude and confidence lacked theoretical conceptualisations. Self-efficacy has provided that theoretical framework for a range of studies subsequently undertaken.

Self-efficacy research

Bandura's (1986, 1997) social learning theory and self-efficacy derivative have provided significant insights into the general behaviour of teachers and have been widely applied to science teaching. The consensus is that teachers with high efficacy beliefs have a high impact on their students' learning. These beliefs also influence the effort teachers make and how long they persist when confronted with obstacles. Terms such as attitude, interest, beliefs, and confidence provide a complex picture of a person's intention to undertake some task. It is beyond the scope of this chapter to explore and differentiate these terms, but these are inter-related constructs, and within the framework of social cognitive theory, they are influenced by higher order self-regulatory abilities. Thus, human actions are a product of the interplay intrapersonal influences and context. Self-efficacy, as a sub-theory of social cognitive theory is conceptualised as a person's assessment of their capability to perform a particular task (Bandura, 1986, 1997). Self-efficacy is not a global trait but a differentiated set of self-beliefs linked to distinct domain. Hence, a teacher's assessment of their capability of teaching science in primary school would define their science teaching self-efficacy within that context. A teacher may have a high sense of self-efficacy in teaching reading to her/his students but a low sense of science teaching self-efficacy.

Bandura argued that a sense of self-efficacy is developed by considering four sources of information: vicarious experiences, performance accomplishments, verbal persuasion, and physiological states. Applied to science teaching, vicarious experiences involve reflecting on the performances of credible models teaching science effectively. Performance accomplishments are derived from successful experiences; thus successfully implementing a science lesson and receiving positive acknowledgment from peers or students. Verbal persuasion occurs when teachers are told they are capable of undertaking the teaching of science. Physiological states reflect the feelings of anxiety or joy in the teaching of science. Tasks that work and generate enthusiasm and excitement should enhance a sense of science teaching self-efficacy. Any one experience quite possibly provides opportunities for several of these sources to impact self-efficacy.

Associated with the construct of self-efficacy as a sense of capability is outcome expectancy. Outcome expectancy is a person's estimation that a given behaviour will lead to certain outcomes. Self-efficacy judgments may be influenced by expected outcomes of performing the task. Tschannen-Moran, Woolfolk Hoy and Hoy (1998) drawing on Bandura's (1986) theoretical framework argued that teachers with high teaching self-efficacy were more willing to try new and varied methods of teaching and had a stronger commitment to teaching. Hoy and Woolfolk (1990) further argued that self-efficacy beliefs can change during pre-service teaching experiences but that in-service teachers are more resistant to change.

In the domain of science teaching, Riggs and Enochs (1990) and Enochs and Riggs (1990) developed the STEBI A (for in-service teachers) and B for (pre-service teachers) instruments which provided a social cognitive theoretical framework to understand the relationship between teacher confidence and practices. The development of a situation-specific instrument represented an important step in the ability of researchers to monitor the status of a teacher's sense of personal teaching efficacy in science and general outcome expectancy with respect to impacts on student learning. The instrument provided a measure of personal science teaching self-efficacy (PSTE) and science teaching outcome expectancy (STOE).

Primary teacher education students often have had limited success in their own science learning, have avoided science, and have naïve ideas about the nature of science (e.g., Palmer, 2002; Watters & Ginns, 1995). Compounded, these experiences and beliefs have consequent effects on their self-efficacy in teaching science and their valuing of science in the primary years of schooling. Although it is tempting to generalize to all primary teachers and pre-service primary teachers, we know that there are teachers who are highly enthusiastic and competent in teaching science. Teachers with high personal teaching self-efficacy have been interested in science for a long time and have a relatively strong background of formal and informal science experiences (de Laat & Watters, 1995).

A number of researchers in the Australian context began to examine whether pre-service teacher education programs or induction programs in the beginning years of school teaching could enhance teacher self-efficacy.

Ginns, Watters and colleagues validated the STEBI instruments for the Australian context (Ginns, Watters, Tulip & Lucas, 1995) and began a series of studies into the attitudes, confidence, and knowledge of pre-service and in-service primary science teachers (Watters & Ginns, 1995, 2000). In a series of longitudinal studies, these researchers concluded that a range of strategies introduced into science methods courses were particularly successful at improving attitudes toward teaching science. Initial findings on pre-service teachers in a foundation science content course showed minimal impact on self-efficacy as measured by STEBI (Ginns et al, 1995). In contrast, the science content course while having no immediate impact on attitudes were valued by students reflecting on curriculum design activities in a subsequent methods course (Watters & Ginns, 1994, 1997a, 1997b, 2000). Watters and Ginns (2000) reached the conclusion that an increase in science content does not automatically result in an increase in efficacy. A conclusion also reached by others (Moore & Watson, 1999; Schoon & Boone, 1998).

Although students valued their science content course, it had been much less effective in developing understanding and confidence than had the science methods course. The authors found that the implementation of student-centered pedagogical practices in pre-service methods courses did result in increased sense of self-efficacy (Watters & Ginns, 2000). This follow-up study, which involved 157 pre-service teachers, had a group effect size of 0.67 on the science teaching self-efficacy scale (PSTE) with no change in outcome expectancy scale (STOE). The instructional strategies implemented by these researchers in a mixed methods study involved collaborative learning workshops, a problem based assignment, and reflective writing; though collaborative learning students were required to adopt metacognitive strategies to evaluate how collaboration was influencing their learning and attitudes to science. The instructional strategies also attempted to recognize that learning in authentic learning environments should simulate experiences that allow students to derive understanding in contexts in which they need to apply that understanding. The tasks employed in the course were modelled on practices that the students would be engaged in as beginning teachers. Most students contrasted a foundation science content course and the science methods course in terms of how the latter course was more effective in increasing their confidence and competence to teach science. Also, supplementary instruction practices using senior pre-service teachers were shown to have a positive effect on student confidence in teaching science (Watters & Ginns, 1997a, 1997b). These findings with pre-service primary teachers were confirmed by Palmer (2002, 2006) and Taylor and Corrigan (2005).

Palmer (2002) adopted an instructional program that involved group inquiry in which the students designed and implemented their own investigations. Also, the sessions involved extensive modelling of teaching strategies similar to those proposed for children. The effect sizes calculated for the pre-test versus immediate post-test means for each scale were 1.74 (PSTE) and 0.89 (STOE). In both cases, the effect sizes were large, above 0.8, indicating that the course had substantially increased personal science teaching efficacy beliefs as well as outcome expectancy beliefs. In addition, Palmer collected data after the pre-service teachers had implemented lessons in practice teaching in schools. An emerging theme was how successful they felt that their science lessons were, which consolidated their confidence. His findings were interpreted as being consistent with Bandura's theory on self-efficacy particularly as the experiences contributed to a sense of successful performance. Anomalous finding in relation to STOE requires further exploration.

In Taylor and Corrigan's (2005) qualitative research they explored the effect of an intervention based on self-regulatory skill development with 19 pre-service primary teachers. The pre-service teachers were required to complete a "learning project" that was centered on investigative work in primary science and required students to design activities, carry them out, and relate them to relevant curriculum documents. Students were asked to identify their specific weaknesses and address these as part of the project. Support was provided in workshops to discuss and scaffold the students' projects. Drawing on interview data, the authors claimed that the project appeared to have considerable success as students came to recognise that they could develop effective science teaching materials.

These studies along with other research conducted in other countries (e.g., Bohning & Hale, 1998; Butts, Koballa & Elliot, 1997; Jarrett, 1999) provided robust evidence that pedagogical practices adopted in pre-service courses which were student centred, involved an inquiry-based approach had positive effects on preservice teachers' confidence.

Clearly, strategies had been identified that enhanced pre-service teachers sense of science teaching efficacy. The questions remained as to how sustainable were these changes and to what extent could experienced teachers be challenged to develop stronger sense of confidence in teaching science. First, a series of small-scale case studies conducted on beginning teachers were examined. In a detailed, three-year longitudinal case study of a single beginning primary teacher, Mulholland and Wallace (2001) mapped the influences that influenced that student's sense of self-efficacy. The research began when Katie, the teacher, was in the second year of a three-year pre-service program. Unfortunately, we are given little information about Katie's prior experiences in science or a baseline measure of her sense of science teaching self-efficacy. Interpreting qualitative data derived from journals, interviews, and observations in a self-efficacy framework, Mulholland argued that few of the conditions necessary to build self-efficacy were evident. Although supervised by experienced teachers, Katie saw little science being taught. Hence, the lack of effective role models further challenged her sense of science teaching self-efficacy. Despite these circumstances, Katie persevered and experienced success in implementing science lessons as judged by student feedback. However, issues of management and class "control" dampened her enthusiasm, and although it appears she became more confident in teaching science, the pedagogical approaches appeared less student centered or innovative than she initially adopted. In this case, achieving mastery experiences of science teaching was an important source of self-efficacy for Katie. Additionally, the scaffolding presumably provided by the researcher contributed to her sense of confidence and obligation to persist.

In a similar study, Appleton and Kindt (2002) explored the experiences of nine beginning teachers in a rural setting. These teachers had performed well in pre-service courses and were considered as competent and enthusiastic beginning teachers. As is normal in this jurisdiction, schools implemented state syllabus documents in ways that were negotiated within individual schools. The culture of the schools where these students were located devalued the teaching of science; therefore the beginning teachers tended to choose science activities that were manageable, conformed to school expectations as communicated by colleagues, and perceived as safe in that they "worked". The activities often meant library research or discussions. Appleton and Kindt argued that: "In one sense this allowed them professional responsibility as curriculum designers but in another it enabled them to evade, if desired, regular teaching of science" (p. 51). Also, they speculated that beginning teachers who were less confident in teaching science would teach those topics that appealed to them whilst they developed basic teaching skills. Thus, they built up an image of self as teacher-based around those preferred subjects, but not science. The importance of collegial support and a school culture that valued science was clearly exposed in this research.

An alternative approach to addressing confidence, again focussed on pre-service teachers, involved collaboration between professional scientists and engineers and university teacher education staff involved in early years education (Howitt, 2010). The project aimed to address the lack of early childhood science resources through developing, implementing, and evaluating various science modules. The project team consisted of five teacher educators and five science/engineering academics.

Teacher educators and science/engineering academics collaboratively developed various science modules and then team taught aspects of these in the early childhood science education workshops.

A further challenge was the general reluctance of in-service teachers to engage in science teaching. Although there were many highly effective and enthusiastic primary science teachers, there was general evidence that science was not uniformly taught across all classes. The importance of early experiences with science, collegial support, and the culture of the school as a vicarious experience emerged in two intensive qualitative studies by de Laat and Watters (1995) and Peers, Diezmann and Watters (2003).

Confronted with a new science program, de Laat and Watters (1995) adopted an ethnographic approach to investigate teachers' attitudes and practices during the adoption of this program. The school of approximately 1000 students was in low socio-economic area with de Laat a deputy principal responsible for leading the introduction of the new science program. Science teaching was initially characterised as lacking coherence with little continuity through the grades. In many classes, the focus was on decontextualized, conceptual understanding with few opportunities for children to engage in real-world problem solving. It was evident that, apart from those few teachers who exhibited a "passion" for science, far too many staff reduced science teaching to a string of unrelated activities derived from prescribed sourcebooks. The school had a strong emphasis on numeracy and literacy to ensure that the students could reach minimum standards. Given this context, a profile of the 37 teachers in the school was undertaken. The STEBI A instrument was used to select teachers (five at each level) who expressed the highest and lowest levels of personal science teaching efficacy. On the PSTE scale scores ranged from 33 to 62. Those teachers with the highest self-efficacy scores, although not necessarily successful at school science themselves, all had an interest in science in several cases stemming from family history. For example, April, one of the teachers in this study, revealed that her father was an engineer and her brother won a science bursary to university. All five had studied formal science during their own schooling and had opportunities to implement or explore science out of school in family situations. Good teachers, role models, or successful employment episodes played a positive part in their positive experiences of science. In contrast, those teachers with low self-efficacy had limited science experiences at school or in the case of one of the teachers attributed her lack of interest to the treatment in science she received in high school because she was a girl.

In a follow-up to this study, Peers, Diezmann and Watters (2003) examined the development of the program through the experiences of one teacher. Notwithstanding constraining issues of time and resources, an experienced teacher reformed his approach to teaching science with support from colleagues. His colleagues shared their experiences, critiqued his approaches, and provided models of practice aligned with the reformed approaches being implemented. He also acknowledged the importance of seeing his students engage enthusiastically with science as a factor in his willingness to abandon his old ways of teaching science and adopt more constructivist-aligned approaches. Thus both vicarious experiences and a sense of success contributed to his increased personal self-efficacy, but one might also speculate that it increased the science teaching self-efficacy of the teaching staff as a whole. Over a five-year period, the level of interest and support for the teaching of science had grown substantially. Thus given the support and conditions that Bandura would argue that impact self-efficacy positively, teachers were capable of engaging more proactively with science and enhancing student outcomes in science. The importance of the school environment as influential on teachers' self-efficacy was further revealed in a study by McKinnon (2010). She noted particularly the effect on outcome expectancy arguing that collaboration with peers enhanced the teachers' confidence that desired outcomes could be achieved.

Despite the strong affirmation of Bandura's theory of self-efficacy as a powerful theory to explain teachers' self-efficacy, researchers continue to seek deeper understandings of primary teachers' hesitancy in teaching science. Mansfield and Woods-McConney (2012) argued that efficacy for teaching primary science is still not completely understood. They have attempted to explore those features of the school and classroom context that influence teacher efficacy providing further confirmation of the importance of successful experiences, vicarious experiences, and enjoyment.

An important contributor to self-efficacy is vicarious experiences. Seeing colleagues or other teachers implementing science can enhance self-efficacy. The use of multimedia was explored in two studies. The first study involved the influence of a television broadcast program on the teaching of science (Watters & Ginns, 1997). The format of the program involved two studio-based teachers and a studio-host leading discussions, implementing activities and reviewing past television broadcasts. The script was supported by video-clips of everyday examples of concepts and processes which were used by the studio-teachers to link the content to real world applications. Interactive engagement with children was achieved through telephone communication with selected schools during each television broadcast. The focus of the study was Anna, an experienced year 4 teacher whose style of teaching was traditional and teacher centred. She was selected on the advice of the principal who describe- her as one of his better science teachers. Anna's pretest STEBI scores were indicative of a high level of science teaching self-efficacy. Anna's enthusiasm about science was in marked contrast to Katie, her year 4 teaching colleague, who despite a substantial background in high school science, which she did not enjoy, avoided teaching science and expressed apprehension about the prospect of having to teach any science at all. Katie's science teaching self-efficacy (PSTE) score was 15 points (almost 2SD) lower than Anna's. It is likely that Katie's beliefs and concerns underpinned the arrangement; whereby Anna assumed the responsibility of teaching science to both year 4 classes. At the end of the eight weeks, Anna's PSTE had dropped moderately (effect size ~.5) and her outcome expectancy had increased substantially (effect size ~.7). The qualitative data indicated that Anna's teaching practices had changed significantly. She employed cooperative group work and active learning activities. The decline in self-efficacy may have been a reflection of her coming to terms with a new way of teaching. The positive influence of watching other teachers teaching science was highlighted in her reflections on the program. Of particular note was that she singled out her observation of the excitement and level of engagement that her students showed in the program, an experience that might have been influential in the change in outcome expectancy scores.

A second study involved the implementation of a multimedia, self-paced, teaching resource depicting the teaching of science in primary classes (Watters & Diezmann, 2003, 2007). A six-component model highlighted the role of: working scientifically, student learning, learning in science, teaching strategies, learning environment, and content. It was developed to guide the production of the multimedia materials. Qualitative data drawn from 100 experienced teachers and 300 pre-service teachers revealed the value of the multimedia material as a vicarious learning experience, the extent that multimedia can demystify science teaching, and the impact on learning outcomes of a multimedia-supported strategy.

Other approaches that have attempted to address issues of beliefs and motivation to teach science in the primary school have involved partnerships with mentors and interactive science centres. Mentors potentially provided strong vicarious experiences. One approach explored the effectiveness of experienced teachers as mentors of primary science (Hudson & Skamp, 2002). It has been assumed that effective mentors for beginning teachers or pre-service teachers in practice teaching situations would develop confidence to teach science. Effective persuasion is one of the contributors to building a sense of self-efficacy. In a study of pre-service teachers being supported by practitioners as mentors, less than a third of the mentors addressed the issue of science teaching "anxiety". Some 83% of participating student teachers also reported that their mentors did not model the teaching of science. The researchers concluded that mentors may not instil positive attitudes toward teaching science.

Interactive science centres were identified by participants, both in-service and pre-service, in McKinnon's (2010) study as a source of inspiration and training. Learning in informal situations, particularly science museums, can generate a great deal of fun and be a source of self-efficacy through its physiological response by engaging participants in active learning. They also provide a resource of ideas that teachers can apply in their classrooms.

Summary

In this section, I have attempted to synthesise the research on primary and early years science teacher efficacy. Defining the constructs of teacher attitudes and beliefs has not been featured strongly in the Australian research. Additionally, much of the research has involved qualitative case study with some surveying. The contexts of teaching science vary widely and comparisons between studies are difficult. For pre-service teachers, each teacher education institution implements courses that reflect their capacity and directions. Primary teacher education programs need to meet certain state teacher accreditations requirements, but these are not specific at the subject level and vary across states. Most universities offer some methods courses, but how disciplinary content is infused varies considerably. What emerges from the research is that personal self-efficacy in teaching science is malleable, and where effective pedagogies are implemented that specifically address issues of confidence, positive changes are evident. Substantial, longitudinal studies are absent, and there is limited research that explores long term changes in teacher efficacy that might occur through major professional development initiatives.

The four sources of efficacy (vicarious experiences, performance accomplishments, verbal persuasion, and physiological states) have contributed to building teachers' sense of science teaching self-efficacy in varying degrees. Successful engagement in learning how to teach science and a better awareness of the conceptual and pedagogical content knowledge needed to teach science has impacted pre-service teachers' efficacy through mastery experiences, vicarious experiences, performance accomplishments, verbal persuasion, and physiological states (e.g., Mansfield & Woods-McConney, 2012; Watters & Ginns, 1995). Similarly, although less convincingly, evidence exists that in-service teachers, having engaged in dedicated science teaching project, have increased their sense of efficacy or confidence through successful experiences (Palmer, 2011). The contribution of resources such as the Australian Academy of Science Primary Connections program (AAS, 2003) has been a prime factor in providing the necessary support and professional development to facilitate mastery. For in-service and pre-service teachers and the collective culture of schools and the exposure to models of good practice have provided vicarious experiences through which teachers have seen credible models implement effective science (de Laat & Watters, 1995; Mansfield & Woods-McConney, 2012). The capacity to talk with colleagues and to have strong leadership in the school focussed on science teaching and the provision of appropriate resources features in many of the studies. The assessment of the role of physiological responses is less clear. Certainly, many teachers have commented in qualitative studies that seeing students engaging in science and enjoying the experience has contributed to their commitment to teach science. Although teachers comment that they enjoy the opportunities to engage students in practical, hands-on work, they are nevertheless concerned with management, loss of control, lack of physical resources, noise, and a feeling of inadequacy when students ask questions.

POLICY RESPONSES

Three significant policy issues have emerged to address the teaching of science in Australia. Seminal to the emergence of these policies was the watershed report of Goodrum, Hackling and Rennie (2001). First, recommendations emerging from their review of the status of science teaching were a number of initiatives around professional development and resourcing. A follow-up to the report was the development of a national action plan. Second, in parallel was the production of a package of resources mapped against the national curriculum and acknowledging the issues confronting the teaching of science in primary schools. Third was the conclusion of the development of a national curriculum.

A major policy initiative resulting from the review of the status of science teaching was The Australian Government Quality Teacher Programme (AGQTP). The government considered this program as a flagship initiative to fund professional development of teachers in a range of areas including science. The project spanned 2004 to 2013, and enabled state education systems to devise programs to meet priority areas especially primary science. A federally funded multi-million dollar Australian School Innovation in Science, Technology and Mathematics [ASISTM] project targeted school based initiatives. These projects targeted upper primary and junior high school and were intended to foster school based projects supported by partnerships with universities or professional associations to enhance student engagement. The scheme lasted three years, but no substantial review of its effectiveness or contribution to teacher efficacy or student engagement has been reported.

The Australian Government commissioned a National Action Plan for Australian School Science Education 2008-2012 (Goodrum & Rennie, 2007). This document written by two of the authors of the first major review of science teaching in Australia (Goodrum, Hackling & Rennie, 2001) was intended as a follow-up and report card on developments between 2001 and 2006. The National Action Plan incorporated recommendations for a range of issues including curriculum, assessment, and teacher training. In relation to teaching, it highlighted the continuing deficit in primary teacher science content knowledge and limited pre-service teacher education to address teacher content knowledge. The plan argued that "Better provision of professional learning and incentives for teachers are required to enable them to maintain their content knowledge of contemporary science, and to improve their pedagogical and pedagogical content knowledge, particularly those inquiry-based pedagogical strategies that develop scientific literacy" (p. 20). It also advocated the use of "excellent teachers of science" (p. 20) as mentors in professional development programs. There was little in the plan specifically to address teacher confidence or self-efficacy; although the research would suggest that the provision of sustained professional development delivered in appropriate ways would be successful in raising levels of self-efficacy.

The disposition to teaching science as an integrated subject and the focus on literacy as a core curriculum emphasis led to the development of the Primary Connections project by the Australian Academy of Science. This project deliberately set out to produce resources to integrate literacy and science on the assumption that teachers are required to teach literacy and, by providing resources that linked science and literacy, science would be taught. The project has provided teachers with print resources based on a sequence of activities mapped into a constructivist framework modelled on the 5Es teaching and learning model (Bybee et al. 1989). The resource contained an extensive collection of text resources all designed for early years and primary teachers with limited confidence and knowledge in science teaching. The materials were supported by a professional development program and a team of consultants. In an evaluation of the program, Hackling and Prain (2007) found evidence that most students stated that they enjoyed learning in science, were rarely bored, and perceived they were learning interesting things. Students from Primary Connections classes were significantly more curious during science lessons and learned interesting things in science than students from comparison classes.

After 20 years in the making, the Australian Government in collaboration with the various state Education Departments adopted a national curriculum. The introduction of this curriculum was phased over several years. Significantly, the first phase targeted English, Science, Mathematics and History. The Australian curriculum purports to address student needs by focussing on the interests and aspirations of students in the primary and junior high school years. The challenge that the curriculum designers provide is the belief that the new curriculum will enable teachers to have the freedom to exercise their professional judgments in the design and implementation of science experiences. Although numeracy and literacy are priority areas with a national assessment strategy being adopted from 2008 to test students in Years 3, 5, 7 and 9, science literacy is also being nationally assessed. In 2009, a sample of approximately five percent of Australian students in year 6 was assessed on scientific literacy. This assessment documents students' capabilities to apply broad conceptual understandings of science to make sense of the world, understand natural phenomena and interpret media reports about scientific issues. It also includes asking investigable questions, conducting investigations, collecting and interpreting data and making decisions (Australian Curriculum Assessment Reporting Authority [ACARA], 2011). Notwithstanding, concerns expressed about high stakes national testing (Lingard, 2010) and the inclusion of science in a national assessment program indicates the priority that science is accorded nationally. Hence, how administrators and teachers of science respond to these national agendas is still emerging.

Substantial professional development opportunities (ASISTM, 2004), (AGQTP, 2004), resources (Australian Academy of Science, 2003), a National Action Plan (2007) and reforms of teacher competencies (MCEECDYA, 2010) all are contributing to a renewed focus on science teaching. Fundamental to the success of these initiatives in the classroom will be teachers' conceptual understanding of science, and also their beliefs, attitudes, and confidence to engage with teaching in ways that heighten interest in science among students.

REFERENCES

ACARA. (2011). *National assessment program, science literacy.* Retrieved October 20, 2012, from http://www.nap.edu.au/NAP_Sample_Assessments/About_each_domain/Science_Literacy/index.html

Australian Curriculum Assessment and Reporting Authority [ACARA]. (2012). *The shape of the Australian curriculum, version 3.* Sydney, NSW: Australian Curriculum, Assessment and Reporting

Authority. Retrieved September 3, 2012, from http://www.acara.edu.au/verve/_resources/The_ Shape_of_the_Australian_Curriculum_V3.pdf

Appleton, K. (1992). Discipline knowledge and confidence to teach science: self-perceptions of primary teacher education students. *Research in Science Education, 22*(1), 11–19. doi: 10.1007/bf02356874

Appleton, K., & Kindt, I. (2002). Beginning elementary teachers' development as teachers of science. *Journal of Science Teacher Education, 13*(1), 43–61.

Appleton, K., & Symington, D. (1996). Changes in primary science over the past decade: Implications for the research community. *Research in Science Education, 26*(3), 299–316.

Appleton, K., Ginns, I. S., & Watters, J. J. (2000). The development of pre-service elementary science teacher education in Australia. In S. K. Abel (Ed.), *Science teacher education: An international perspective* (pp. 9–29). Dordrecht, The Netherlands: Kluwer Academic Publishers.

Aubusson, P., & Webb, C. (1992). Teacher beliefs about learning and teaching in primary science and technology. *Research in Science Education, 22*(1), 20–29. doi: 10.1007/bf02356875

Australian Academy of Science [AAS]. (2003). *Primary Connections: Linking science with literacy.* Canberra, ACT: Australian Academy of Science.

Australian Academy of Technological Sciences and Engineering [ATSE]. (2002). *The teaching of science and technology in Australian primary schools: A cause for concern.* Parkville, VIC: Australian Academy of Technological Sciences and Engineering.

Australian Education Council. (1989). *The Hobart declaration on schooling.* Hobart, TAS: Australian Education Council.

Baker, R. (1994). Teaching science in primary schools: What knowledge do teachers need? *Research in Science Education, 24*(1), 31–40. doi: 10.1007/bf02356326

Bandura, A. (1986). *Social foundations of thought and action: A social cognitive theory.* Englewood Cliffs, NJ: Prentice Hall.

Bandura, A. (1997). *Self-efficacy: The exercise of control.* New York, NY: Freeman.

Bandura, A. (2001). Social cognitive theory: An agentic perspective. *Annual Review of Psychology, 52*(1), 1–26.

Biddulph, F., & Osborne, R. (1984). *Making sense of our world: An interactive teaching approach: Handbook.* Hamilton, NZ: Science Education Research Unit, University of Waikato.

Bohning, G., & Hale, L. (1998). Images of self-confidence and the change-of-career prospective elementary science teacher. *Journal of Elementary Science Education, 10*, 39–59.

Butts, D. P., Koballa Jr., T. R., & Elliot, T. D. (1997). Does participating in an undergraduate elementary science methods course make a difference? *Journal of Elementary Science Education, 9*, 1–17.

Bybee, R., Buchwald, C., Crissman, S., Heil, D., Kuebis, P, Matsumoto, C., & McInerney, J. (1989). *Science and technology education for the elementary years: Frameworks for curriculum and instruction.* Andover, MA: The National Center for Improving Science Education. (ERIC ED314237)

Curriculum Corporation (1994). *A statement on science for Australian schools.* Melbourne, VIC: Curriculum Corporation.

de Laat, J., & Watters, J. J. (1995). Science teaching self-efficacy in a primary school: A case study. *Research in Science Education, 25*(4), 453–464.

Department of Employment Education and Training [DEET]. (1989). *The discipline review of teacher education in mathematics and science (1). Report and recommendations.* Canberra, ACT: Australian Government Publishing Services.

Dobson, I. R. (2012). *Unhealthy science? University natural and physical sciences 2002 to 2009/10.* Canberra, ACT: Office of the Chief Scientist.

Enochs, L., & Riggs, I. (1990). Further development of an elementary science teaching efficacy instrument: A pre-service elementary scale. *School Science and Mathematics, 90*, 694–706.

Fensham, P. J. (1993). Reflections on science for all. In L. Whitelegg (Ed.), *Challenges and opportunities in science education.* Milton Keynes: The Open University.

Fensham, P., & Northfield, J. R. (1993). Pre-service science teacher education: An obvious but difficult arena for research. *Studies in Science Education, 22*, 67–84.

Ginns, I., Watters, J. J., Tulip, D. F., & Lucas, K. B. (1995). Changes in pre-service elementary teachers' sense of efficacy in teaching science. *School Science and Mathematics, 95*(8), 394–400.

Goodrum, D., Druhan, A., & Abbs, J. (2011). *The status and quality of year 11 and 12 science in Australian schools. Report prepared for the Office of the Chief Scientist*. Canberra: Australian Academy of Science. Retrieved October 21, 2012, from http://www.science.org.au/reports/documents/Year-1112-Report-Final.pdf

Goodrum, D., Hackling, M., & Rennie, L. (2000). *The status and quality of the teaching and learning of science in Australian schools*. Report to the Department of Education, Training and Youth Affairs. Canberra: DETYA.

Goodrum, D., & Rennie, L. (2007). *Australian School Science Education National Action Plan, 2008–2012 Volume 1: The national action plan*. Canberra, ACT.

Hardy, T., Bearlin, M., & Kirkwood, V. (1990). Outcomes of the primary and early childhood science and technology education project at the University of Canberra. *Research in Science Education, 20*(1), 142–151. doi: 10.1007/bf02620489

Harlen, W. (1997). Primary teachers' understanding in science and its impact in the classroom. *Research in Science Education, 27*(3), 323–337.

Harlen, W., & Qualter, A. (2004). *The teaching of science in primary schools*. London: David Fulton.

Hattie, J. (2009). *Visible learning for teachers: A synthesis of over 800 meta-analyses relating to achievement*. Oxford: Routledge.

Howitt, C. (2010). *Science for early childhood teacher education students (ECTES): Collaboration between teacher educators, scientists and engineers*. Surry Hills, NSW: Australian Teaching and Learning Council.

Hoy, W. K., & Woolfolk, A. E. (1990). Socialization of student teachers. *American Education Research Journal, 27*, 279–300.

Hudson, P., & Scamp, K. (2002). Mentoring pre-service teachers of primary science. *Electronic Journal of Science Education, 7*(1). Retrieved October 2012, from http://ejse.southwestern.edu/article/view/7692/5459

Jarrett, O. S. (1999). Science interest and confidence among pre-service elementary teachers. *Journal of Elementary Science Education, 11*, 47–57.

Lingard, B. (2010). Policy borrowing, policy learning: Testing times in Australian schooling. *Critical Studies in Education, 51*(2), 129–147. doi: 10.1080/17508481003731026

Loucks-Horsley, S., & Matsumoto, C. (1999). Research on professional development for teachers of mathematics and science: The state of the scene. *School Science and Mathematics, 99*(5), 258–271.

Mansfield, C. F., & Woods-McConney, A. (2012). I didn't always perceive myself as a science person: Examining efficacy for primary science teaching. *Australian Journal of Teacher Education, 37*(10), 37–52.

McDonald, C. V. (2010). The influence of explicit nature of science and argumentation instruction on preservice primary teachers' views of nature of science. *Journal of Research in Science Teaching, 47*(9), 1137–1164. doi: 10.1002/tea.20377

MCEECDYA. (2010). *National professional standards (Draft)*. Carlton South, VIC: Ministerial Council for Education, Early Childhood Development and Youth Affairs. Retrieved October 21, 2012, from http://www.mceecdya.edu.au/verve/_resources/NPST-DRAFT_National_Professional_Standards_for_Teachers.pdf

McKinnon, M. C. (2010). *Influences on the science teaching self efficacy beliefs of Australian primary school teachers*. Dissertation, Australian National University, Canberra, ACT.

Moore, J. J., & Watson, S. B. (1999). Contributors to the decision of elementary education majors to choose science as an academic concentration. *Journal of Elementary Science Education, 11*, 37–46.

Mulholland, J., & Wallace, J. (2001). Teacher induction and elementary science teaching: enhancing self-efficacy. *Teaching and Teacher Education, 17*(2), 243–261. doi: 10.1016/s0742-051x(00)00054-8

Organisation for Educational and Cultural Development (OECD). (2010). *PISA 2009 results: what students know and can do; student performance in reading, mathematics and science*. Paris: OECD Publishing. Retrieved September 1, 2012, from http://dx.doi.org/10.1787/9789264091450-en

Osborne, J. (2007). Science education for the twenty first century. *Eurasia Journal of Mathematics, Science & Technology Education, 3*(3), 173–184.

Osborne, J., & Simon, S. (1996). Primary science: Past and future directions. *Studies in Science Education, 26*, 99–147.

Palmer, D. (2006). Durability of changes in self-efficacy of pre-service primary teachers. *International Journal of Science Education, 28*(6), 655–671.

Palmer, D. H. (2002). Factors contributing to attitude exchange amongst pre-service elementary teachers. *Science Education, 86*(1), 122–138. doi: 10.1002/sce.10007

Peers, C. E., Diezmann, C. M., & Watters, J. J. (2003). Supports and concerns for teacher professional growth during the implementation of a science curriculum innovation. *Research in Science Education, 33*(1), 89–110. doi: 10.1023/a:1023685113218

Riggs, I. M., & Enochs, L. G. (1990). Toward the development of an elementary teacher's science teaching efficacy belief instrument. *Science Education, 74*(6), 625–637.

Schoon, K. J., & Boone, W. J. (1998). Self-efficacy and alternative conceptions of science of pre-service elementary teachers. *Science Education, 82*, 553–568.

Shulman, L. S. (1987). Knowledge and teaching: Foundations of the new reform. *Harvard Educational Review, 57*, 1–21.

Symington, D. (1974). Why so little primary science? Australian Science Teachers' Journal, 20(1), 57–62.

Taylor, N., & Corrigan, G. (2005). Empowerment and confidence: Pre-service teachers learning to teach science through a program of self-regulated learning. *Canadian Journal of Science, Mathematics and Technology Education, 5*(1), 41–60. doi: 10.1080/14926150509556643

Tschannen-Moran, M., Woolfolk Hoy, A., & Hoy, W. K. (1998). Teacher efficacy: Its meaning and measure. *Review of Education Research, 68*, 202–248.

Venville, G., Wallace, J., & Louden, W. (1998). The primary science teacher-leader project. *Research in Science Education, 28*(2), 199–217.

Watters, J. J., & Diezmann, C. M. (2003, April 7–10) Windows into a science classroom: Making science relevant through multimedia resourcing. In *ICASE World Conference 2003 on Science and Technology Education*, Penang, Malaysia.

Watters, J. J., & Diezmann, C. M. (2007). Multimedia resources to bridge the praxis gap: Modeling practice in elementary science education, *18*(3), 349–375. doi: 10.1007/s10972-007-9051-x

Watters, J. J., & Ginns, I. S. (1995). *Origins of and changes in pre-service teachers' science teaching self-efficacy.* Paper presented at the Annual Meeting of the National Association for Research in Science Teaching, San Francisco.

Watters, J. J., & Ginns, I. S. (1997a). *Impact of course and program design features on the preparation of pre-service elementary science teachers.* Paper presented at the Annual Meeting of the National Association of Research in Science Teaching, Chicago, IL. (ERIC ED 408267).

Watters, J. J., & Ginns, I. S. (1997b). *Peer assisted learning: Impact on self-efficacy and achievement.* Paper presented at the Annual meeting of the American Educational Research Association, Chicago.

Watters, J. J., & Ginns, I. S. (2000). Developing motivation to teach elementary science: Effect of collaborative and authentic learning practices in pre-service education. *Journal of Science Teacher Education, 11*(4), 301–321.

AFFILIATION

James J Watters
Faculty of Education,
Queensland University of Technology

INDEX

CPSIA information can be obtained at www.ICGtesting.com
Printed in the USA
LVOW05*0157151014

408808LV00006B/40/P

9 789462 095564